Elmar Eder

Relative Complexities of First Order Calculi

Artificial Intelligence

Künstliche Intelligenz

edited by Wolfgang Bibel and Walther von Hahn

Artificial Intelligence aims for an understanding and the technical realization of intelligent behaviour.
The books of this series are meant to cover topics from the areas of knowledge processing, knowledge representation, expert systems, communication of knowledge (language, images, speach, etc.), AI machinery as well as languages, models of biological systems, and cognitive modelling.

In English:

Automated Theorem Proving
by Wolfgang Bibel

Parallelism in Logic
by Franz Kurfeß

Relative Complexities of First Order Calculi
by Elmar Eder

In German:

Die Wissensrepräsentationssprache OPS 5
by Reinhard Krickhahn and Bernd Radig

Prolog
by Ralf Cordes, Rudolf Kurse, Horst Langendörfer and Heinrich Rust

LISP
by Rüdiger Esser and Elisabeth Feldmar

Logische Grundlagen der Künstlichen Intelligenz
by Michael R. Genesereth and Nils J. Nilsson

Wissensbasierte Echtzeitplanung
by Jürgen Dorn

Modulare Regelprogrammierung
by Siegfried Bocionek

Automatisierung von Terminierungsbeweisen
by Christoph Walther

Logische und Funktionale Programmierung
by Ulrich Furbach

Schließen bei unsicherem Wissen in der Künstlichen Intelligenz
by Léa Sombé

Elmar Eder

Relative Complexities of First Order Calculi

AMS Subject Classification: 68 T 27, 68 T 15, 03 B 35, 03 B 10, 03 F 20

Verlag Vieweg · P. O. Box 58 29 · D-6200 Wiesbaden · FR Germany

Vieweg is a subsidiary company of the Bertelsmann Publishing International.

Softcover reprint of the hardcover 1st edition 1992

Cover design: L. Markgraf, Wiesbaden

ISBN-13: 978-3-528-05122-8 e-ISBN-13: 978-3-322-84222-0
DOI: 10.1007/978-3-322-84222-0

Acknowledgements
The first ideas for this book developed during research at the Technische Universität München with Wolfgang Bibel. Major advances were made during a short stay granted to me at Duke University in Durham, North Carolina, and during stays as a visiting professor at various German Universities. The ideas and the main results of Chapter 4 developed during a stay granted to me at the University of British Columbia (UBC) in Vancouver, Canada. Chapter 3 is an outcome of my stay as a visiting professor at the Technische Hochschule Darmstadt. Espescially fruitful were discussions with my colleagues at the ICOT in Tokyo where I was granted a stay as a visiting researcher.

I want to thank Armin B. Cremers for his support. I thank Kurt Schütte for teaching me the foundations of mathematical logic. I owe special thanks to Wolfgang Bibel from whome I learned a lot about automated theorem proving, and through whose initiative and support my visits to many renowned universities and institutes were possible. I want to thank my colleagues here and abroad vor valuable discussions.

Contents

Introduction

In this paper, a comparison is made of several proof calculi in terms of the lengths of shortest proofs for some given formula of first order predicate logic with function symbols. In particular, we address the question whether, given two calculi, any derivation in one of them can be simulated in the other in polynomial time. The analogous question for propositional logic has been intensively studied by various authors because of its implications for complexity theory. And it seems there has not been as much endeavour in this field in first order logic as there has been in propositional logic. On the other hand, for most of the practical applications of logic, a powerful tool such as the language of first order logic is needed.

The main interest of this investigation lies in the calculi most frequently used in automated theorem proving, the resolution calculus, and analytic calculi such as the tableau calculus and the connection method. In automated theorem proving there are two important aspects of complexity. In order to have a good theorem proving system, we must first have some calculus in which we can express our derivations in concise form. And second, there must be an efficient search strategy. This book deals mainly with the first aspect which is a necessary condition for the second since the length of a shortest proof always also gives a lower bound to the complexity of any strategy. Moreover, the search space increases exponentially with the length of a proof. Especially in interactive theorem proving systems where the search is partly left to the user's intuition, the naturalness and conciseness of the representation of proofs is among the main issues. This motivates our interest in lengths of shortest proofs in a given first order calculus, and in its ability to simulate proof techniques which are expressible in other proof calculi.

In Chapter 1, several calculi for first order predicate logic are described. The calculi considered here are Robinson's resolution calculus, Bibel's connection method, Beth's and Smullyan's method of tableaux, Gentzen's sequent calculus, Gentzen's natural deduction calculus, and a Frege-Hilbert calculus, and the consolution calculus. Since the connection method is not a fixed calculus, but rather a design philosophy for calculi, I had to make a choice for one particular calculus. The calculus chosen contains most of the features relevant to the lengths of shortest proofs that are included in the various connection calculi described so far. On the other hand, it is simple enough to be easily seen to be sound. One feature that is not contained in this calculus is splitting. Since it is not obvious how splitting can be done when it is nested, there was no connection calculus available in which splitting is formally defined in a completely satisfactory way. The connection structure calculus introduced in Chapter 4 can be viewed as such a calculus, however, since it allows to simulate splitting in any form in which it is known to be correct; and it is, in fact, a generalization of splitting. It is shown that the consolution calculus can simulate resolution as well as the connection calculus in a very simple and streight-forward way.

In Chapter 2, most of these calculi are compared with each other. After a brief review of some of the major results known so far in this field, a polynomial time transformation of a formula to clausal form, called its definitional form, is described which is needed for comparing calculi for full first order logic with calculi for clausal form logic. Measures are given and investigated for measuring the complexities of resolution refutations. In the course of a resolution refutation, the cardinalities of the clauses may grow exponentially; so the number of resolution steps is not always an adequate measure of complexity. The first simulation result presented here is a step by step simulation of the chosen connection calculus by resolution. The converse is not possible. There is a class of formulas for which the length of the shortest proof in the connection calculus grows superpolynomially with respect to the length of their shortest resolution proof. The two main reasons for this inefficiency are the unability of the connection calculus considered here to forget variables that are not needed any more, and the fact that the use of lemmata is possible only in a very restricted way in this calculus. Then variants of the tableau calculus are considered, and it is shown that the version of the connection method considered here, and a suitable variant of the tableau calculus, are essentially identical.

In Chapter 3, Tseitin's extension rule is generalized to first order logic, and it is proved that Tseitin's and Reckhow's result, stating that resolution can simulate the cut-free sequent calculus at polynomial cost, and that resolution with extension can simulate the sequent calculus with cut at polynomial cost, carries over from propositional logic to first order logic. More precisely, the calculus of extended definitional resolution is introduced. A derivation in this calculus consists of a transformation of a given arbitrary formula of first order predicate logic to its definitional form, and a subsequent derivation of this definitional form by resolution and extension. It is then proved that extended definitional resolution can p-simulate the sequent calculus, and that definitional resolution (without extension) can p-simulate the sequent calculus without the cut rule. Moreover, the simulation maps tree derivations in the sequent calculus to tree derivations in extended definitional resolution. From Gentzen's simulation results, it then follows that extended definitional resolution can also p-simulate natural deduction and the Frege-Hilbert calculus. Moreover, it is proved that, even when these calculi are augmented by the feature of definition, they can be p-simulated by extended definitional resolution.

In Chapter 4, the *connection structure calculus* is introduced. The connection structure calculus is a calculus which is derived from the connection method. Instead of the usual concept of unification it uses the concept of a *unifier set* which presents a way to incorporate the process of forgetting of variables in the concept of unifiers and of unification. It is shown how each resolution proof can be expanded (at exponential cost) to a connection proof and how such connection proofs can be represented more efficiently as dags, called *connection structures*. In particular, multiple nested use of lemmata does not cause the exponential explosion in this representation that it does in the connection calculus described in Section 1.3. We give a formal definition of connection structures, and we give some theorems holding for connection structures. A proof calculus, the *connection structure calculus*, is presented that is based on connection structures. To this end, annotations are introduced for each node of the dag. These annotations are needed to keep track of the unification process during derivation and of the set of paths still to be checked for complementarity at each instant of time (in the language of the connection method). In resolution, the object corresponding to this set of paths is the resolvent of the considered resolution step. It is shown how splitting can be done in the connection structure calculus.

Chapter 1

Calculi for First Order Logic

In this chapter we give a brief description of a few proof calculi for first order predicate logic with function symbols which are suitable for automated theorem proving and most of which have been used in actual implementations by various authors. Section 1.1 gives some basic concepts of first order predicate logic and of automated theorem proving, and some general remarks on the question of the suitability of calculi for the automation of reasoning. The following sections 1.2, 1.3, 1.5, 1.6, 1.7, and 1.8 contain descriptions of resolution, the connection method, the method of tableaux, the sequent calculus, the natural deduction calculus, and a Frege-Hilbert calculus, respectively.

1.1 Basic Concepts and General Remarks

1.1.1 First Order Predicate Logic

This book deals with the problem of automated theorem proving in first order predicate logic with function symbols but without equality. We shall not give a full introduction to first order predicate logic here, but rather refer the reader to the literature. There are many books giving an introduction to the field. Let us just mention the books by Hermes [Her76], Hilbert and Ackermann [HA72], Hilbert and Bernays [HB68], Shoenfield [Sho67], Smullyan [Smu71], and Stegmüller and Varga [SVvK84]. Most of the calculi described in this chapter are treated in the book of Bläsius and Bürckert [BB87].

The alphabet of the language we are concerned with here consists of the propositional connectives $\neg$, $\vee$ and $\wedge$, the quantifiers $\exists$ and $\forall$, a finite or infinite number of function symbols and predicate symbols—zero or more of arity n for every nonnegative integer n—, a countably infinite number of variables and, as auxiliary symbols, the parentheses and the comma, '(', ')' and ','. The nullary function symbols are also called *constants*. In some sections we shall also use the propositional connectives $\rightarrow$ and $\leftrightarrow$.

The concepts of a *term* and of a *formula* are defined inductively as follows.

1. Every variable is a term.

2. If f is an n-ary function symbol and $t_1, \ldots, t_n$ are terms then $f(t_1, \ldots, t_n)$ is a term.

3. If P is an n-ary predicate symbol and $t_1, \ldots, t_n$ are terms then $P(t_1, \ldots, t_n)$ is a formula, called an *atomic formula*.

4. If F is a formula then $\neg F$ is a formula.

5. If F and G are formulas then $(F \vee G)$ and $(F \wedge G)$ are formulas.

6. If F is a formula and x is a variable then $\exists x F$ and $\forall x F$ are formulas.

We adopt the usual rules of omitting parentheses.

Now, the problem of automated theorem proving is to find an automatic procedure for deciding, at least for some subset of the set of all formulas, whether a given formula is valid, i.e. whether it is true with respect to every interpretation of the function symbols and predicate symbols. Since the problem of validity of a formula of first order predicate logic is undecidable[1], we cannot expect to get a correct answer from the system for all formulas. The best we can expect is to obtain the answer "yes" for every valid formula and "no" for some subset of the set of non-valid formulas, but then there are always formulas for which the procedure does not terminate.

Various calculi have been developed by logicians since the late nineteenth century for a formalization of first order predicate logic. Some of the landmarks of this development are Gottlob Frege's Begriffsschrift [Fre79] and the introduction of Frege-Hilbert calculi, Gerhard Gentzen's sequent calculus, and his calculus of natural deduction in [Gen35]. All these calculi are adequate for the purpose they were designed for, namely as a tool for the mathematician and for a formal study of what it means to make logical inferences. All of them describe the same set of derivable formulas. All logical inferences used by mathematicians in proofs of mathematical theorems can be translated to any of these calculi as has been shown (for one such calculus) by Whitehead and Russel in their Principia Mathematica [WR13].[2]

In automated theorem proving, however, there are some more requirements that have to be met by a proof calculus. One such requirement is that the search should have a small branching factor. At each instant of time during the proof process there should be only a small number of choices what to do next. In Frege-Hilbert calculi, for example, this requirement is not fulfilled since there you start

[1] Of course, this undecidability can be asserted only under the assumption that Church's thesis (see [Chu36]) holds. It is in that case an immediate consequence of Theorem X in Kurt Gödel's 1931 paper [Göd31] and was shown by Alonzo Church in [Chu36].

[2] Actually, the Begriffsschrift allows quantification also over predicate symbols, and it allows to use formulas and predicate symbols as terms. This leads to inconsistencies in connection with the use of the concept of definitions as Bertrand Russel has shown. These inconsistencies are removed in the Principia Mathematica by introduction of a typed logic which allows to build higher order formulas as well as first order formulas. But by a suitable formalization of set theory in first order logic, higher order formulas can be avoided.

from the axioms and derive from them the goal formula that you want to prove, and there is an infinite number of possible axioms you can choose at every step.

This requirement has led to calculi that use a top down or goal-directed process of proof search, starting from the goal formula and going back to the axioms. Furthermore, it has led to the introduction of the concept of unification which is central to automated theorem proving.

In Sections 1.2, 1.3 and 1.5 of this book some of these calculi are described, and Section 2 contains a comparative study of these calculi.

1.1.2 Substitutions and Unification

Here are the basic definitions for the concepts of substitution and unification, and the notations used for them in this book.

Definition 1.1.1
By $\mathcal{V}$ we denote the set of variables. A *substitution* is a function σ mapping variables to terms such that the set $\{x \in \mathcal{V} \mid \sigma(x) \neq x\}$, called the *domain* of σ, is finite. We denote substitutions by lower case Greek characters. If $x_1, \ldots, x_n$ are pairwise distinct variables and $t_1, \ldots, t_n$ are terms then $\{x_1 \leftarrow t_1, \ldots, x_n \leftarrow t_n\}$ denotes the substitution σ defined by $\sigma(x_j) \stackrel{\text{def}}{=} t_j$ for $j = 1, \ldots, n$, and $\sigma(x) \stackrel{\text{def}}{=} x$ for $x \notin \{x_1, \ldots, x_n\}$. If t is a term or a quantifier-free formula or a set of terms or a set of quantifier-free formulas then we denote by $\sigma(t)$ or by σt the result of replacing each occurrence of a variable x in t by the term $\sigma(x)$.

Definition 1.1.2
If σ and τ are substitutions then the *composition* $\sigma\tau$ of σ and τ is the substitution defined by $(\sigma\tau)x \stackrel{\text{def}}{=} \sigma(\tau x)$ for $x \in \mathcal{V}$. A substitution σ is said to be *more general* than a substitution τ if there exists a substitution ρ such that $\tau = \rho\sigma$.

Definition 1.1.3
Let S be a set of terms or of atomic formulas and let σ be a substitution. Then σ is said to be a *unifier* of S if $\sigma s = \sigma t$ for all $s, t \in S$. The substitution σ is said to be a *most general unifier* (mgu) of S if σ is a unifier of S and σ is more general than any unifier of S. It is denoted by $\text{mgu}(S)$.

We shall also consider simultaneous unifiers of a set S of sets of terms (or atomic formulas), meaning a substitution that is a unifier of each element of S. A most general unifier of S is then, again, a unifier of S that is more general than any unifier of S. It will be denoted by $\text{mgu}(S)$. Moreover, we shall also consider negated atomic formulas. If A and B are atomic formulas and A' is A or $\neg A$, and B' is B or $\neg B$, then by a unifier of A' and B' we mean a unifier of A and B, i.e., we ignore negation signs.

Definition 1.1.4
A *renaming substitution* or *renaming of variables* or *permutation* is a substitution which is a bijection on the set of variables. If T is a term or a quantifier-free formula or a set of quantifier-free formulas and π is a renaming substitution then we say that πT is a *variant* of T.

1.2 Resolution

Most automatic theorem proving systems that have been implemented so far use the *resolution calculus*. Resolution has been introduced by J. A. Robinson in [Rob65] as a method for automated theorem proving. An introduction to automated theorem proving, and especially to resolution can be found in the books of Chang and Lee [CL73], Loveland [Lov78], Wos, Overbeek, Lusk and Boyle [WOLB84], and others. See also Siekmann and Wrightson [SW83].

Resolution is a refutation calculus, i.e., a formula F is not proved directly to be valid, but its negation $\neg F$ is proved to be unsatisfiable. Since a formula F is valid iff $\neg F$ is unsatisfiable it is reasonable to proceed this way. In resolution the language of first order predicate logic is restricted to a proper subset, the language of *clausal form* predicate logic.

Definition 1.2.1
A *literal* is an atomic formula or the negation of an atomic formula. A *positive literal* is an atomic formula. Its *sign* is $+$. A *negative literal* is a negation of an atomic formula. Its *sign* is $-$. By a *clause formula* we mean a formula of the form

$$\forall x_1 \ldots \forall x_k (L_1 \vee \cdots \vee L_n)$$

where $L_1, \ldots, L_n$ are literals and $x_1, \ldots, x_k$ are the variables occurring in the literals $L_1, \ldots, L_n$. A *clause* is a finite set of literals. A formula is said to be in *clausal form*, or in *conjunctive normal form* if it has the form

$$C_1 \wedge \cdots \wedge C_n$$

where $C_1, \ldots, C_n$ are clause formulas.

Here we assume any fixed associativity of $\vee$ and of $\wedge$, say we choose them to be right associative. Because of the associativity laws of $\vee$ and $\wedge$ which hold modulo semantic equivalence, it makes no difference whether we use them in a right or left associative way.

Resolution is a refutation calculus for formulas in clausal form, i.e., a calculus for proving unsatisfiability of formulas in clausal form. In the resolution calculus, a clause formula is represented by the set of its literals, and a formula in clausal form is represented by the set of clauses representing its conjuncts. Usually, a clause formula is identified with the clause representing it. So we sometimes use the word 'clause' to mean both, a set of literals, and the universal closure of the disjunction of these literals. Likewise, a formula in clausal form is identified with the set of clauses representing it.

Definition 1.2.2
Let c be a clause. Then we say that a clause d is a *factor* of c if there are two distinct literals K and L in c, both positive or both negative, and if there is a most general unifier σ of K and L such that $d = \sigma c$.

Definition 1.2.3
Let c and d be two clauses. Then we say that a clause e is a *resolvent* of c and d if there are d', K, L' and σ such that the following three conditions hold

1. d' is a variant of d which has no variables in common with c.

2. $K \in c$ and $L' \in d'$ are literals of opposite sign which are unifiable and have σ as a most general unifier.

3. $e = (\sigma c \setminus \{\sigma K\}) \cup (\sigma d' \setminus \{\sigma L'\})$.

The clauses c and d are called the *parent clauses*. If L is the literal of d that L' is a variant of, then the literals K and L are called the *literals resolved upon*.

Definition 1.2.4
Let S be a set of clauses. A *resolution refutation* of S is a finite sequence $(c_1, \ldots, c_n)$ of clauses such that c_n is the empty clause $\Box$ and such that for each $k \in \{1, \ldots, n\}$ at least one of the following three conditions holds.

1. $c_k \in S$

2. There is an $i \in \{1, \ldots, n\}$ such that $i < k$ and c_k is a factor of c_i.

3. There are $i, j \in \{1, \ldots, n\}$ such that $i, j < k$ and c_k is a resolvent of c_i and c_j.

We say that a set of clauses is *refutable* by resolution (or in the resolution calculus) iff there exists a resolution refutation of it. We shall speak of *resolution steps* (or just *resolutions*) and of *factorization steps* (or just *factorizations*) in a resolution refutation. Instead of 'factorization' we shall also say '*explicit factorization*'. We say that a resolution step contains an *implicit factorization* if at least one of the following two conditions holds.

1. There are two distinct literals $H, J \in c$ or $H, J \in d'$ such that $\sigma H \equiv \sigma J$.

2. There are literals $H \in c$ and $J \in d'$ not resolved upon in the resolution step such that $\sigma H \equiv \sigma J$.

The following soundness and completeness theorem holds for the resolution calculus.

Proposition 1.2.1
A set of clauses is refutable in the resolution calculus iff the set of corresponding clause formulas is unsatisfiable.

In order to obtain a resolution refutation of a formula this formula must be in clausal form. Still, resolution can be used for theorem proving in full first order logic. There is a way of transforming any formula F of first order predicate logic to a formula F' in conjunctive normal form, called its *conjunctive Skolem normal form*, such that F is satisfiable iff F' is satisfiable. So if you want to prove a formula of full first order predicate logic to be valid using the resolution calculus then what you have to do is the following. First negate F. Then transform $\neg F$ to its conjunctive Skolem normal form, say G, and try to refute G in the resolution calculus. If G is refutable in the resolution calculus then F is valid. A discussion of the problems concerning the time and space complexity of a transformation to clausal form will be given in Section 2.2.

1.3 The Connection Method

In this section we give a brief introduction to the connection method which was introduced by Wolfgang Bibel as a method of automated theorem proving. For a detailed exposition of this method see his book Automated Theorem Proving [Bib87][3].

The connection method can directly handle arbitrary formulas of first order predicate logic. They need not be in clausal form as in resolution. However, using the transformation given in Section 2.2 of this book, any formula of first order predicate logic can be transformed in a natural way to some formula in clausal form. The transformation given there does not have the disadvantages of the transformation usually used for this purpose, namely exponential time complexity, exponential increase of the length of the formula, and disruption of the structure of the given formula. Moreover, the search space is the same for both formulas when using the connection method. So we shall restrict our attention to the clausal form versions of the connection method in this book in order to be able to make a direct comparison to resolution.

Another difference between resolution and the connection method is that resolution is formulated as a method for proving the unsatisfiability of a formula in conjunctive normal form whereas the connection method is formulated as a method for proving the validity of a formula in disjunctive normal form.

Definition 1.3.1
In the connection method, by a *clause formula* we mean a formula of the form

$$\exists x_1 \dots \exists x_k (L_1 \wedge \dots \wedge L_n)$$

where $L_1, \dots, L_n$ are literals and $x_1, \dots, x_k$ are the variables occurring in the literals $L_1, \dots, L_n$. In the connection method, a formula is said to be in *clausal form* if it has the form

$$C_1 \vee \dots \vee C_n$$

where $C_1, \dots, C_n$ are clause formulas. In this case we also say that the formula is in *disjunctive normal form*.

[3]Here we use a slightly different notation from his.

Definition 1.3.2
The *dual* of a formula F is the formula obtained from F by replacing all occurrences of $\vee$ by $\wedge$ and vice versa, and all occurrences of $\exists$ by $\forall$ and vice versa.

The dual of a clause formula for the connection method is a clause formula for the resolution method, and vice versa. A formula is in clausal form for the resolution method iff its dual is in clausal form for the connection method, and vice versa.

Deriving the validity of a formula in disjunctive normal form and refuting a formula in conjunctive normal form, are dual to each other. Instead of viewing resolution as a method of refuting a formula F you can just as well consider it as a method of deriving the validity of the dual of F,[4] and instead of viewing the connection method as a method of deriving the validity of a formula F you can just as well consider it as a method of refuting the dual of F. In fact, in [And81] Peter Andrews describes a method for theorem proving—which is essentially the dual to the connection method—through what he calls 'acceptable matings'. An acceptable mating is the dual of what is called a 'spanning unifiable set of connections' in the connection method.

In [Bib83], Wolfgang Bibel has compared the connection method and the method of matings. He points out that the connection calculus he presented is more efficient than Andrews' calculus of matings, due to more efficient coding and proof search and to a feature called splitting he introduced to the connection method.

If S is a finite set of clauses then we shall speak of validity or of unsatisfiability of S, both meaning the same thing, namely that the formula F represented by S in the connection method is valid. Since the formula G represented by S in the resolution method is the dual of F this is equivalent to saying that G is unsatisfiable. Moreover, a set of clauses is refutable in the resolution method iff it is derivable in the connection method.

Whereas resolution presents us with just one fixed calculus for which we can then choose some suitable strategy, the connection method can be viewed as a method for designing proof calculi. This makes it in principle impossible to compare "the resolution method" and "the connection method". We have to choose some existing version(s) of the connection method for the comparison with resolution. In this book, we choose the basic version of the connection method, with

[4]You might object and say that the resolution calculus is a calculus for proving the unsatisfiability of a whole set S of clauses, and S may be infinite (of course, in general S will be recursively enumerable, at least relative to some oracle whose materialization may, for example, be the user in an interactive system). But then the unsatisfiability of S is equivalent to the proposition that, for all interpretations, at least one formula of T is true where T consists of the duals of all formulas of S. And this is exactly what the connection method establishes if it is given an infinite set T of input clauses that has this property, and a similar statement holds of other affirmative theorem proving calculi. This is a consequence of the compactness theorem (Gödel [Göd30], Theorem X). From now on we shall, for simplicity, assume that we are dealing only with finite sets of clauses.

just one additional feature, namely backward factorization. This calculus is simple enough to be provably correct.

In [Bib83] and [Bib87], Wolfgang Bibel presents connection calculi with more advanced features such as a treatment of arbitrary formulas (not necessarily in clausal form and not skolemized), some enhancements resulting in a pruning of the search tree, and splitting. Except for splitting, these features do not affect the numbers of proof steps of shortest proofs although they can reduce the search tree considerably. The length of a single proof step changes only by a factor which is greater than the reciprocal of the maximum number of symbols in a clause. For an efficient transformation to clausal form see Section 2.2.

1.3.1 The Connection Method in Propositional Logic

In this section we consider formulas of propositional logic in *disjunctive normal form*, i.e., disjunctions of conjunctions of literals. Each disjunct is represented as a *clause*, the set of literals it is a conjunction of. The formula itself is represented as the set of its clauses, called the *matrix*. The clauses can be represented graphically as vertical columns which are put together horizontally into the matrix.

For example, the formula $\neg U \vee (U \wedge V) \vee \neg W$ has the matrix representation $\{\{\neg U\}, \{U, V\}, \neg W\}$, or, graphically,

$$\begin{matrix} & U & \\ \neg U & & \neg W \\ & V & \end{matrix}$$

A *path* through a matrix (or through the corresponding formula) is a set of literals, one from each clause. A *connection* in a matrix or in a formula is an unordered pair of occurrences of complementary literals X and $\neg X$ in the matrix or formula. A path is said to be *complementary* iff it contains a connection, i.e., iff there are two literals in the path which are complementary to each other. A set C of connections in a matrix or formula is said to be *spanning* iff every path through this matrix or formula contains at least one connection of the set C. A matrix is said to be *complementary* iff each of its paths is complementary, or, equivalently, if the set of its connections is spanning.

In the example above there are just one connection $\{\neg U, U\}$ and two paths $\{\neg U, U, \neg W\}$ and $\{\neg U, V, \neg W\}$ the first of which is complementary and the second is not. The set of connections is not spanning in this example. So the matrix is not complementary.

The connection method is based on the following

Proposition 1.3.1
A formula is valid iff its matrix representation is complementary.

The idea of the connection method is to systematically check the paths through the given matrix M for complementarity. The procedure starts by choosing arbitrarily one clause c of the matrix as a starting clause, then choosing one literal L of this clause and marking the remaining literals of this clause as *unsolved subgoals*. Then all paths that pass through the chosen literal are checked for complementarity. Then the same is done in turn for all paths passing through any of the unsolved subgoals of the starting clause, each time removing the corresponding literal from the set of unsolved subgoals. In order to check all paths passing through some specific literal K of the first clause for complementarity we choose a second clause from which we again choose some literal L and mark the remaining literals as unsolved subgoals, and so on. So we obtain, at each instant of time, a sequence of literals $(K, L, \ldots)$, called the *active path* p.

For every set p of literals and every matrix M let us denote by $\text{compl}(p, M)$ the statement

> "Every path through M is complementary or contains a literal which is complementary to some literal of p."

Then M is complementary iff $\text{compl}(\emptyset, M)$ is true.

Proposition 1.3.2
Let p be a set of literals and let M be a matrix. Then the following two propositions hold.

1. *If $M = \emptyset$ then* $\text{compl}(p, M)$ *is false.*
2. *If $M \neq \emptyset$ and $c \in M$ is arbitrary then* $\text{compl}(p, M)$ *is true iff for all $L \in c$ at least one of the following two conditions holds.*
   (a) L is complementary to some literal of p.
   (b) $\text{compl}(p \cup \{L\}, M \setminus \{c\})$ *is true.*

This proposition can be considered as a declarative formulation of a non-deterministic and inherently parallel algorithm for deciding whether a given matrix is complementary or not. It is non-deterministic due to the liberty concerning the choice of the clause c, and the different choices of literals L in the clause c introduce an and-parallelism in the case that M is to be proved complementary, and an or-parallelism in the case that M is to be proved non-complementary.

Let us now give a description of the connection calculus that is close to the original one given in W. Bibel's book [Bib87]. It is, essentially, an iterative and sequential formulation of the algorithm rendered by Proposition 1.3.2. The removal of recursion from the algorithm must result in one or more stacks containing information concerning the steps that have yet to be done at each instant of time.

In summary, we have, at each instant of time:

- the finite sequence $(c_1, \ldots, c_{k+1})$ of chosen clauses of M
- the finite sequence $D = (d_1, \ldots, d_{k+1})$ of the sets of *unsolved subgoals* in the corresponding clauses
- the finite sequence $p = (L_1, \ldots, L_k)$ of chosen literals, called the *active path*.

We have $d_j \subseteq c_j$ for $j = 1,\ldots,k+1$, and $L_j \in c_j \setminus d_j$ for $j = 1,\ldots,k$. In order to formulate the inference rules of the connection method it is superfluous to explicitly mention the sequence of chosen clauses. Instead we define:

Definition 1.3.3
A *structured matrix* is a triple (M, p, D) where M is a matrix, p is a sequence of k literals and D is a sequence of $k+1$ clauses, and k is some nonnegative integer.

Before we come to the inference rules of the calculus let us have a look at the differences between our version of the connection calculus and the original version given by Wolfgang Bibel in [Bib87]. In [Bib87], the active path p is represented by a function γ that maps each clause c_j intersecting the active path p to the literal L_j such that $p \cap c_j = \{L_j\}$. The set of unsolved subgoals is represented there by a function β mapping each unsolved subgoal (each element of $\bigcup D$) to 1 and all other literals of the concerned clauses to 0. The set $\{c_1, \ldots, c_{k+1}\}$ of clauses of which the clauses of D are taken as subsets is denoted by F_1 there, and the order of the clauses of F_1 is represented by a function α mapping each such clause c_j to the positive integer j.

Except for these notational differences there are only three points here that differ from W. Bibel's formulation.

1. We allow multiple use of one clause of M. Disallowing multiple use of clauses, as Bibel does it for propositional logic, is a good strategy, but it does not effect the length of a shortest proof of a formula, which is the domain of interest of this book. In fact, every derivation containing a multiple use of a clause can be shortened to one without multiple uses of clauses. Moreover, this restriction is not contained any more in the connection method for first order logic since it would render the calculus incomplete.

2. In an extension step (described below) Bibel has the additional restriction that there must be a literal in the clause extended to that is complementary to some literal of the active path. Again, this restriction can be viewed as a search strategy that is of no relevance to the lengths of shortest derivations that we study in this book.

3. In addition to the inference rules of extension and truncation described below there is a rule of separation in Bibel's book. It is applied when the other two rules are not applicable. It consists of throwing away clauses that can be shown not to be able to contribute to the derivation, and of retrying a derivation with other clauses. Separation is necessary for the strong completeness of Bibel's calculus. This is due to the restriction in the rule of extension and does not apply to the calculus presented here which is strongly complete under certain fairness assumptions, even without the separation rule. Strong completeness is lost in first order predicate logic in Bibel's calculus as well as in the version presented here, even with the separation rule. The separation rule has no influence on the lengths of shortest proofs since it consists

of discarding clauses. Every derivation of a formula with separation can be transformed to a derivation of the same formula without separation by just omitting all deduction steps that lie between a choice of a clause c in an extension step and the discarding of the same clause c in a separation step.

Now we come to the inference rules of the calculus. In the basic version of the connection method there are two inference rules, extension and truncation. In the refinement that we consider here we have, in addition, the rule of (backward) factorization.

Definition 1.3.4
We say a structured matrix (M, q, E) is obtained from a structured matrix (M, p, D) by an *extension step*[5],

$$(M, p, D) \vdash_{\text{ext}} (M, q, E),$$

if $p = (L_1, \ldots, L_k)$, $D = (d_1, \ldots, d_{k+1})$, $L_{k+1} \in d_{k+1}$, $q = (L_1, \ldots, L_{k+1})$, $E = (d_1, \ldots, d_k, d_{k+1} \setminus \{L_{k+1}\}, e)$, and e is obtained from a clause c of M by deleting literals that are complementary to literals of q.

In other words, an extension step consists in choosing a literal L_{k+1} from the last clause of the sequence D, extending the active path p through L_{k+1} to obtain the new active path q, removing the literal L_{k+1} from the set of unsolved subgoals, choosing a clause c—called the *clause extended to*—from M and adding those literals of M that are not complementary to any literals of the path q to the set of unsolved subgoals.

Of course, any extensions of the active path q through literals of $c \setminus e$ are complementary. These are exactly the paths that are being checked for complementarity at this extension step.

Definition 1.3.5
We say a structured matrix (M, q, E) is obtained from a structured matrix (M, p, D) by a *truncation step*,

$$(M, p, D) \vdash_{\text{tru}} (M, q, E),$$

if $p = (L_1, \ldots, L_k)$, $D = (d_1, \ldots, d_{j+1}, \emptyset, \ldots, \emptyset)$ with $k - j$ occurrences of $\emptyset$, $q = (L_1, \ldots, L_j)$, and $E = (d_1, \ldots, d_{j+1})$, where $0 \leq j < k$.

In other words, truncation means retracting the active path from its last $k - j$ literals if there are no unsolved subgoals in the last $k - j$ chosen clauses.

During a truncation step no paths are checked for complementarity. A truncation step serves merely as a preparation for the next extension step, and to determine the next path to be checked for complementarity.

[5] This concept of extension is not related to the extension rule introduced in Section 3!

Although these two rules are already sufficient to yield a sound and complete calculus for propositional clausal form logic we shall add, as a third rule, the *factorization rule* which allows in some cases to avoid to solve a literal more than once that occurs in the matrix several times. There are cases where the complementarity of some path p_1 can be inferred from the complementarity of some other path p_2. In such a case the complementarity check of p_1 can be omitted if it is guaranteed that the complementarity of p_2 will be checked in a later stage of the derivation. This is called (backward) factorization. For more details see Bibel's book [Bib87].

Definition 1.3.6
We say a structured matrix (M, q, E) is obtained from a structured matrix (M, p, D) by a *factorization step*,

$$(M, p, D) \vdash_{\text{fac}} (M, q, E),$$

if $p = q = (L_1, \ldots, L_k)$, $D = (d_1, \ldots, d_k, d_{k+1})$, L is a literal with $L \in d_{k+1}$, and $E = (d_1, \ldots, d_k, d_{k+1} \setminus \{L\})$, and there is a literal K in one of the sets $d_1, \ldots, d_k$ which is identical with L.

We call the unordered pair $\{K, L\}$ the *factorization link chosen* in the factorization step. To be precise, if we speak of a connection or factorization link $\{K, L\}$, then K and L are understood to be occurrences of literals rather than literals. So, if L and L' are different occurrences of the same literal in a matrix, and if $\{K, L\}$ is a connection or factorization link in this matrix, then $\{K, L'\}$ is considered to be a different connection or factorization link. We shall omit the word 'occurrence' wherever there is no ambiguity. For details concerning occurrences of literals in the connection method, we refer the reader to [Bib87].

Definition 1.3.7
A *connection derivation* or *connection proof* of a matrix M is a finite sequence $(S_0, \ldots, S_r)$ of structured matrices where $S_0 = (M, (), (c))$ for some clause $c \in M$, $S_r = (M, (), (\emptyset))$, and, for each $s = 1, \ldots, r$, we have $S_{s-1} \vdash_{\text{ext}} S_s$ or $S_{s-1} \vdash_{\text{fac}} S_s$ or $S_{s-1} \vdash_{\text{tru}} S_s$. M is said to be *derivable* in the connection method iff there exists a connection derivation of M.

Then the following soundness and completeness holds.

Proposition 1.3.3
A formula is valid iff the corresponding matrix is derivable in the connection method.

Definition 1.3.8
A set C of connections together with a set F of factorization links in a matrix M is *spanning* if each path through M contains two literal occurrences K_0 and L_0 such that there are non-negative integers k and l and literal occurrences $K_1, \ldots, K_k$ and $L_1, \ldots, L_l$ in M such that $\{K_k, L_l\} \in C$ and $\{K_{i-1}, K_i\} \in F$ for $i = 1, \ldots, k$, and $\{L_{j-1}, L_j\} \in F$ for $j = 1, \ldots, l$.

Proposition 1.3.4
If a set of connections together with a set of factorization links in a matrix is spanning then every formula represented by this matrix is valid.

1.3.2 The Connection Method in First Order Predicate Logic

In this section we consider the connection method for formulas of first order predicate logic in *disjunctive normal form*, i.e., disjunctions of existential closures of conjunctions of literals. Again, each disjunct is represented as a *clause*, the set of its literals. The formula is represented as a *matrix* which is the set of its clauses.

For example the formula $\neg P(a) \vee \exists x(P(x) \wedge \neg P(f(x))) \vee P(f(f(a)))$ is represented as the matrix $\{\{\neg P(a)\}, \{P(x), \neg P(f(x))\}, \{P(f(f(x)))\}\}$, or, graphically,

$$\begin{array}{ccc} & P(x) & \\ \neg P(a) & & P(f(f(a))) \\ & \neg P(f(x)) & \end{array}$$

With this example we see two problems that were not present in propositional logic. First, there are two connections in this example, $\{\neg P(a), P(x)\}$ and $\{\neg P(f(x)), P(f(f(a)))\}$. Now, for each of these connections, it does not hold that one of the two connected literals is the negation of the other. Instead, in order to make one of the connections complementary we have first to apply a substitution to the two involved clauses. For this substitution we take the most general unifier of the two connected literals. Second, in this example the connections are not simultaneously unifiable. The remedy we have to take for this is to use two variants of the second clause, thus obtaining the matrix

$$\begin{array}{cccc} & P(x_1) & P(x_2) & \\ \neg P(a) & & & P(f(f(a))) \\ & \neg P(f(x_1)) & \neg P(f(x_2)) & \end{array}$$

Now, the substitution $\{x_1 \leftarrow a, x_2 \leftarrow f(a)\}$ is a most general unifier of the spanning set of connections

$$\{\{\neg P(a), P(x_1)\}, \{\neg P(f(x_1)), P(x_2)\}, \{\neg P(f(x_2)), P(f(f(a)))\}\}.$$

Here we see another problem that was not present in propositional logic. Namely, even after generating two variants of the second clause, the set of all connections is not unifiable. We have to choose a suitable spanning subset of the set of all connections that is unifiable. This means also that, in contrast to propositional logic, backtracking is necessary in an implementation of the calculus, because some chosen connection might not have been the right one to choose.

Definition 1.3.9
A *connected matrix* is a set of clauses together with a set of connections. It is called *complementary* if its set of connections is spanning and unifiable. A matrix is said to be *complementary matrix* if there is a unifiable spanning set of connections in a finite set of variants of its clauses.

The connection method for first order predicate logic is based on the following proposition which is a consequence of Herbrand's Theorem.

Proposition 1.3.5
A formula is valid iff its matrix is complementary.

Definition 1.3.10
A *structured matrix* is a quadruple (M, p, D, σ) where M is a matrix, p is a sequence of k literals, k is some nonnegative integer, D is a sequence of $k+1$ clauses, and σ is a substitution.

This concept of a structured matrix is very similar to the corresponding concept in propositional logic. Instead of a triple (M, p, D) we have a quadruple (M, p, D, σ) where the *current substitution* σ is the most general simultaneous unifier of all connections chosen so far in extension steps and of all factorization links chosen so far in factorization steps.

Definition 1.3.11
We say a structured matrix (M, q, E, τ) is obtained from a structured matrix (M, p, D, σ) by an *extension step*,

$$(M, p, D, \sigma) \vdash_{\mathbf{ext}} (M, q, E, \tau),$$

if there are $k, L_1, \ldots, L_{k+1}, d_1, \ldots, d_{k+1}, e, c, \mathcal{C}$ and ρ such that the following conditions hold.

1. $p = (L_1, \ldots, L_k)$ and $D = (d_1, \ldots, d_{k+1})$ and $L_{k+1} \in d_{k+1}$.
2. $q = (L_1, \ldots, L_{k+1})$ and $E = (d_1, \ldots, d_k, d_{k+1} \setminus \{L_{k+1}\}, e)$.
3. c is a variant of a clause of M.
4. $\mathcal{C}$ is a set of connections between literal occurrences of q and literal occurrences of c.
5. e is the set of literal occurrences of c which are not elements of connections in $\mathcal{C}$.
6. ρ is a most general unifier of $\sigma\mathcal{C}$, and $\tau = \rho\sigma$.

There are a few differences from an extension step in propositional logic. In an extension step in first order logic, only a subset of the set of all possible connections is chosen. We shall refer to the connections of $\mathcal{C}$ as the *connections chosen* in that extension step. The action of choosing a connection we call a *choice of a connection* done in the extension step. Starting from the empty substitution, the current substitution σ is merged at each extension or factorization step (see below) with the most general unifier of the connections or of the factorization link chosen at that step, in order to obtain the new current substitution.

Definition 1.3.12
We say a structured matrix (M, q, E, σ) is obtained from a structured matrix (M, p, D, σ) by a *truncation step*,

$$(M, p, D, \sigma) \vdash_{\mathbf{tru}} (M, q, E, \sigma),$$

if $p = (L_1, \ldots, L_k)$, $D = (d_1, \ldots, d_{j+1}, \emptyset, \ldots, \emptyset)$ with $k - j$ occurrences of $\emptyset$, $q = (L_1, \ldots, L_j)$, and $E = (d_1, \ldots, d_{j+1})$, where $0 \leq j < k$.

Definition 1.3.13
We say a structured matrix (M, q, E, τ) is obtained from a structured matrix (M, p, D, σ) by a *factorization step*,

$$(M, p, D, \sigma) \vdash_{\mathbf{fac}} (M, q, E, \tau),$$

if there are $L_1, \ldots, L_k, d_1, \ldots, d_{k+1}, L, K$ and ρ such that the following conditions hold.

1. $p = q = (L_1, \ldots, L_k)$
2. $D = (d_1, \ldots, d_k, d_{k+1})$
3. $L \in d_{k+1}$
4. $E = (d_1, \ldots, d_k, d_{k+1} \setminus \{L\})$
5. $K \in d_1 \cup \cdots \cup d_k$, and K has the same sign as L.
6. ρ is a unifier of σK and σL, and $\tau = \rho\sigma$.

The concept of a *derivation* is defined similarly as in propositional logic. Again we have a soundness and completeness theorem.

Proposition 1.3.6
A formula is valid iff the corresponding matrix is derivable.

The proof (for the original connection calculus) can be found in [Bib87].

1.3.3 Splitting

The problem of proving a formula can sometimes be splitted in two or more subproblems. Take, for example, the following matrix from [BEF83] and [Ede87].

$$\begin{array}{cc} \neg P(x) & P(y) \\ \neg P(f(x)) & P(f(f(y))) \end{array}$$

This set of clauses has a resolution refutation in 4 resolution steps. Namely, let $c_1 \stackrel{\text{def}}{=} \{\neg P(x), \neg P(f(x))\}$, $c_2 \stackrel{\text{def}}{=} \{P(y), P(f(f(y)))\}$, $d_1 \stackrel{\text{def}}{=} \{\neg P(f(y)), P(y)\}$, $d_2 \stackrel{\text{def}}{=} \{\neg P(f(x))\}$, $d_3 \stackrel{\text{def}}{=} \{P(y)\}$, and $d_4 \stackrel{\text{def}}{=} \square$. Then d_1 is a resolvent of c_1

and c_2, d_2 is a resolvent of c_1 and d_1, d_3 is a resolvent of c_2 and d_2, and d_4 is a resolvent of d_2 and d_3.

Now, the shortest derivation of this matrix in the connection calculus without splitting involves 4 variants of the first clause and 3 variants of the second clause. The corresponding connected matrix is depicted below where parentheses are omitted where this does not lead to ambiguities.

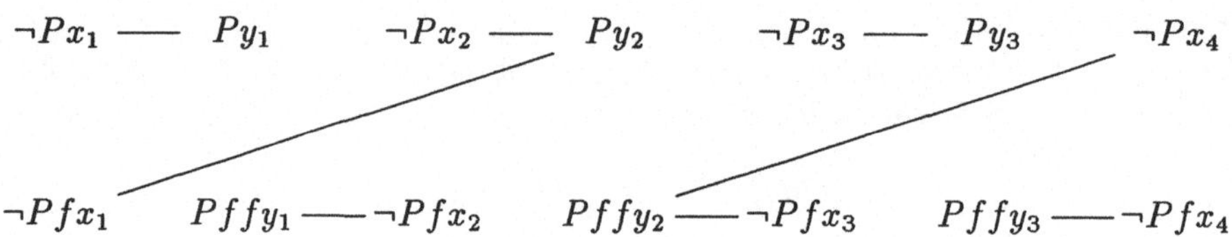

The connections are indicated as straight lines. The connection $\{\neg Pfx_1, Py_2\}$ can be replaced by the factorization link $\{\neg Pfx_1, \neg Px_2\}$, and the connection $\{Pffy_2, \neg Px_4\}$ can be replaced by the factorization link $\{\neg Pfx_3, \neg Px_4\}$. Both of these factorization links are indicated by dotted lines below.

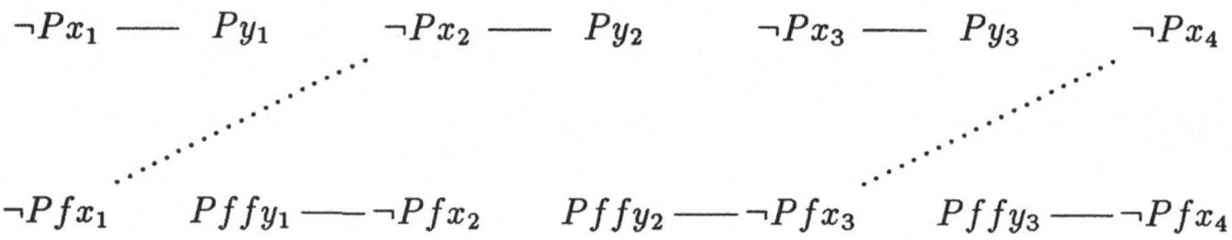

Note that the submatrix consisting of the last three clauses of this connected matrix is a variant of the submatrix

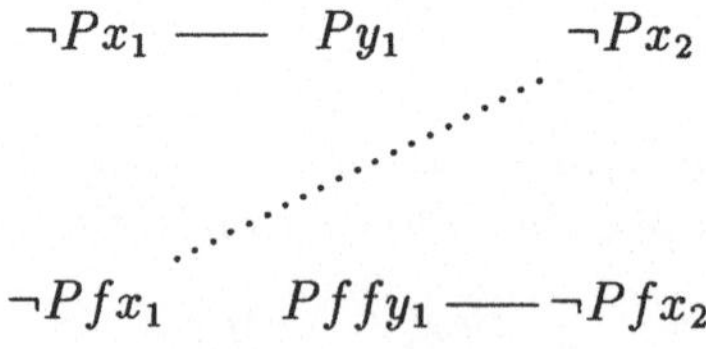

consisting of its first three clauses. If we merge these two submatrices to a single one — identifying x_3 with x_1, y_3 with y_1, and x_4 with x_2 —, we obtain the connected matrix

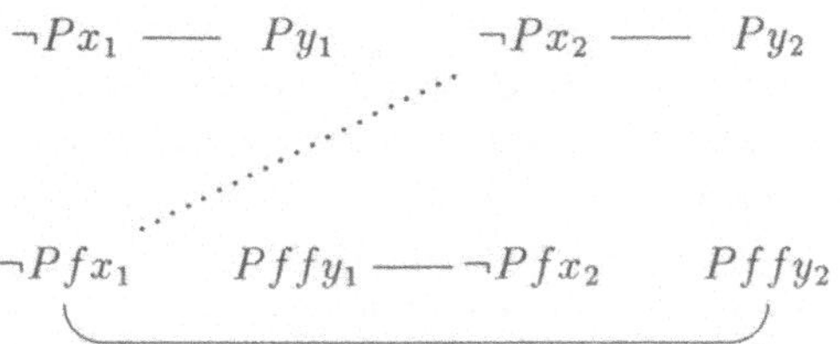

This is still a connected matrix consisting of variants of the input clauses. Its set of connections together with its set of factorization links is still spanning, but it is not unifiable. The non-unifiability becomes more apparent if we replace the connection $\{\neg Pfx_1, Pffy_2\}$ with the connection $\{\neg Px_2, Pffy_2\}$ to obtain the connected matrix shown in Figure 1.1. Then the two connections between the third clause and the last clause are not compatible with each other in the sense that they have no common unifier. This is indicated in Figure 1.1 by the horizontal line separating the two literals in the last clause.

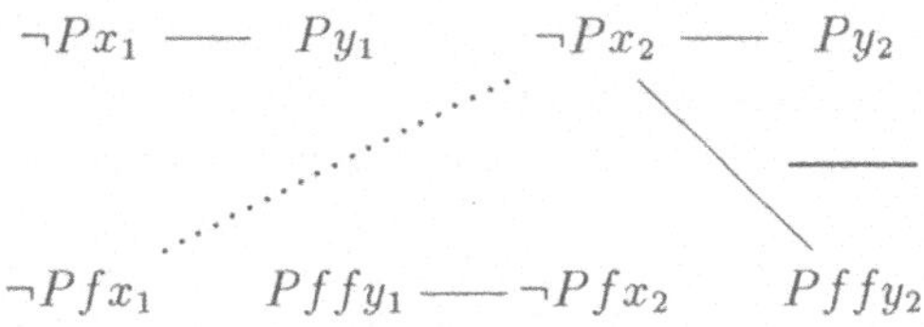

Figure 1.1: A connected matrix with splitting

Definition 1.3.14
We say that a substitution σ is *more general* than a substitution τ *on a set of variables* V if there is a substitution ρ such that $\tau x = \rho\sigma x$ for all $x \in V$.

Definition 1.3.15
A set Σ of substitutions is said to be *compatible* on a set V of variables if there is a substitution τ such that each element of Σ is more general than τ on the set V.

Now, in a derivation in the connection method with splitting, it is not required any more that the set of connections in the resulting connected matrix be compatible on the set of all variables. Rather, the compatibility condition can be weakened to hold only on a subset of the set of all variables, according to the following proposition.

Proposition 1.3.7
Let H be a formula represented by a set S of clauses. Let M be a connected

matrix consisting of a set of variants of clauses of S together with a set C of connections and a set F of factorization links. Assume that C together with F is spanning for M. Let $c \in M$. For each literal occurrence $L \in c$, let C_L be the set of connections of M not involving a literal occurrence of $c \setminus \{L\}$, and let F_L be the set of factorization links of M not involving a literal occurrence of $c \setminus \{L\}$. Assume that, for each $L \in c$, there is a substitution unifying each connection in C_L and each factorization link in F_L. Let σ_L be a most general such substitution, i.e., a most general unifier of $C_L \cup F_L$. Furthermore, we assume that the unifiers σ_L are chosen in such a way, that for different literals $K, L \in c$, the clause $\sigma_K c$ does not have any variables in common with the clause $\sigma_L c$. Such a choice is always possible if an mgu of $C_L \cup F_L$ exists for all $L \in c$. For each $L \in c$ let τ_L be the restriction of σ_L to the set of variables occurring in c. Assume that the set of substitutions $\{\tau_L \mid L \in c\}$ is compatible on the set of variables occurring in c. Then the formula H is valid.

Proof: The idea of the proof is to consider the matrix M_0 consisting of all clauses of M except the clause c. Then, for each literal L of c, a new variant M_L of M_0 is chosen. Let M' be the matrix $\bigcup_{L \in c} M_L \cup \{c\}$. In M' consider the following set of links (connections and factorization links). Take all links inherited from M in each M_L. Furthermore, for each literal $L \in c$, those links from the clause c into the submatrix M_L are taken which are inherited from M. Then this set of links is spanning for M' and unifiable. Thus the formula represented by M is valid.

q.e.d.

In our example in Figure 1.1, the clause c is the last (fourth) clause of the matrix. If K is the literal Py_2 then the set C_K consists of the connections $\{\neg Px_1, Py_1\}$, $\{Pffy_1, \neg Pfx_2\}$ and $\{\neg Px_2, Py_2\}$, and the set F_K consists of the factorization link $\{\neg Pfx_1, \neg Px_2\}$. The substitution σ_K is $\{y_1 \leftarrow x_1, x_2 \leftarrow fx_1, y_2 \leftarrow fx_1\}$ and $\tau_K = \{y_2 \leftarrow fx_1\}$. If L is the literal $Pffy_2$ then the set C_L consists of the connections $\{\neg Px_1, Py_1\}$, $\{Pffy_1, \neg Pfx_2\}$ and $\{\neg Px_2, Pffy_2\}$, and the set F_L consists of the factorization link $\{\neg Pfx_1, \neg Px_2\}$. The substitution σ_L is $\{x_1 \leftarrow fy_2', y_1 \leftarrow fy_2', x_2 \leftarrow ffy_2', y_2 \leftarrow y_2'\}$ and $\tau_L = \{y_2 \leftarrow y_2'\}$. Let $\tau \stackrel{\text{def}}{=} \{y_2 \leftarrow fx_1\}$. Then the hypotheses of Proposition 1.3.7 are fulfilled, and hence the formula is valid. Exactly the same substitution τ is obtained when we have the connection $\{\neg Pfx_1, Pffy_2\}$ instead of $\{\neg Px_2, Pffy_2\}$. Two literals K and L of M which are not in c and which are connected to each other by a factorization link can be regarded equivalent in the sense that any connection $\{J, K\}$ can be replaced by $\{J, L\}$. Note that, in our example, the connected matrix with 7 clauses is obtained back again by the construction in the proof of Proposition 1.3.7.

The idea of splitting is to use the connection method to prove a formula, and, as soon as the set of links ceases to be unifiable, to try to split one of the clauses of the matrix, to weaken the requirements for unifiability of links according to Proposition 1.3.7, and to continue the derivation with this weakened condition. We shall not go into any more details of the concept of splitting here. Rather, we refer the reader to Bibel's book [Bib87]. Note that, if the set $\{\tau_L \mid L \in c\}$ is compatible on a set V of variables, then there is a most general substitution τ on V such that all the substitutions τ_L with $L \in c$ are more general than τ on V. This follows from the fact that the set of substitutions occurring as unifiers can be turned into a complete lattice with respect to the 'more general than'-relation by factoring with respect to a suitable equivalence relation and adding a top element. The construction of the lattice and the proof of this fact are given in [Ede85b]. As a consequence, a nested use of splitting is possible. We shall see in Chapter 4 how this can be achieved by means of connection structures introduced there.

1.4 Consolution

A rather simple but powerful calculus combining the idea of the connection method with the idea of resolution is the *consolution calculus* [Ede91]. As the connection calculus described in the last section, also consolution is based on the concept of paths.

By a *path in* a matrix we mean a subset of a path through a matrix. Thus, a path in a matrix is a set of literals, at most one chosen from each clause. We shall also use the term *partial path* for a path in a matrix. An *extension* of a path p in a matrix M is a path q in M such that $p \subseteq q$. If c is a clause, p and q are finite sets of literals and $\mathcal{P}$ and $\mathcal{Q}$ are finite sets of finite sets of literals then we define

$$\begin{aligned} \mathcal{P}_c &\stackrel{\text{def}}{=} \{\{L\} \mid L \in c\} \\ pq &\stackrel{\text{def}}{=} p \cup q \\ \mathcal{P}\mathcal{Q} &\stackrel{\text{def}}{=} \{pq \mid p \in \mathcal{P} \text{ and } q \in \mathcal{Q}\}. \end{aligned}$$

$\mathcal{P}_c$ is the set of (one-element) paths through the (one-clause) matrix $\{c\}$. If M and N are disjoint matrices, ie., disjoint finite sets of clauses, then pq is a path in $M \cup N$ for every path p in M and for every path q in N. Similarly, $\mathcal{P}\mathcal{Q}$ is a set of paths in $M \cup N$ for every set $\mathcal{P}$ of paths in M and for every set $\mathcal{Q}$ of paths in N. We call the set $\mathcal{P}\mathcal{Q}$ the *product* of the sets $\mathcal{P}$ and $\mathcal{Q}$.

1.4.1 Consolution in propositional logic

The idea behind consolution is the following. In order to prove the validity of a given formula using the connection method, all paths through its matrix have to be checked for complementarity. At each stage of a connection proof a certain set of paths has already been checked for complementarity whereas all the remaining paths still have to be processed this way. In consolution at each stage of a proof process, this remaining set of paths yet to be checked for complementarity is encoded. It would be inefficient, however, to explicitly represent each path of this set since the number of paths in general increases exponentially with the number of clause instances. Instead, this set of paths is coded in the form of a set of partial paths. Each partial path encodes the set of all its extensions through the whole matrix. So one partial path may encode many paths through the matrix. Let us consider an example which for simplicity is taken from propositional logic.

Suppose we want to prove the validity of the formula

$$(P \wedge Q) \vee (\neg P \wedge Q) \vee \neg Q.$$

The following tree is a proof tree of this formula by consolution.

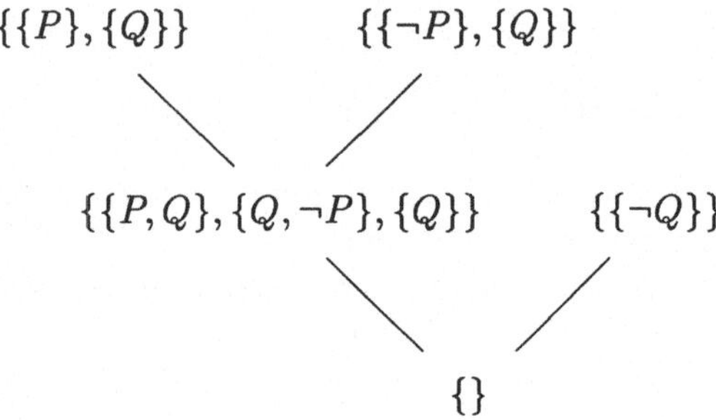

Each node of the tree is marked with a set of (partial) paths in the matrix M

$$\begin{matrix} P & \neg P & \\ & & \neg Q \\ Q & Q & \end{matrix}$$

representing the given formula. A leaf of the proof tree is marked with the set $\mathcal{P}_c$ of all one-element paths through some clause c of the formula, ie. $\mathcal{P}_c = \{\{L\} \mid L \in c\}$. Since every path through the given matrix M is an extension of a path through c for every $c \in M$, the set $\mathcal{P}_c$ encodes the set of all paths through M. Thus the marks $\mathcal{P}_c$ of the leaves reflect the fact that at the beginning of the proof process all paths through M still have to be checked for complementarity.

The inference rule, also called *consolution*, takes the set of partial paths from each premise and combines these sets into a new set, called the *consolvent*. This combination can be regarded to consist of two parts. The first part consists of building the product $\mathcal{PQ}$ of $\mathcal{P}$ and $\mathcal{Q}$. For example, if $\mathcal{P} = \{\{P\},\{Q\}\}$ and $\mathcal{Q} = \{\{\neg P\},\{Q\}\}$ then $\mathcal{PQ} = \{\{P,\neg P\},\{P,Q\},\{Q,\neg P\},\{Q\}\}$. The second part of the inference rule consists in simplifications of $\mathcal{PQ}$. One such simplification is the elimination of complementary paths. To continue the illustration of our example, this means that the complementary path $\{P,\neg P\}$ is removed from $\mathcal{PQ}$. Actually, both parts are to be seen as a single operation, a remark which bears its relevance on the efficiency of implementation. But for the ease of the reader's understanding we will continue to make the distinction.

The proof in our example is completed by applying consolution once again to the result of the previously illustrated step and the remaining leaf $\{\{\neg Q\}\}$. The resulting empty set is the criterion of a successful derivation (as in resolution). To summarize, for a formula F in disjunctive normal form, a *proof* of F is a derivation of the empty set from the sets $\mathcal{P}_c$ $(c \in M)$ with the consolution rule, where M is the set of clauses of F.

While this explains the essentials of the calculus, the following additional details complete its description. Above we have used a single simplification which is elimination of complementary paths. This is all needed for completeness and soundness of the calculus. For efficiency, it is necessary to incorporate at least the following simplification (since otherwise a huge number of paths will quickly be generated in practice). Any path p in $\mathcal{PQ}$ may be replaced by any subset of p. Both simplifications may be applied simultaneously. So in our present example, the first consolvent $\{\{P,Q\},\{Q,\neg P\},\{Q\}\}$ may be further reduced, for instance to $\{\{Q\}\}$. Note that three distinct paths have been replaced by a single one in this case. In general, the number of paths may be reduced considerably this way. If we incorporate this simplification into the derivation shown above, the following proof tree of the same formula results.

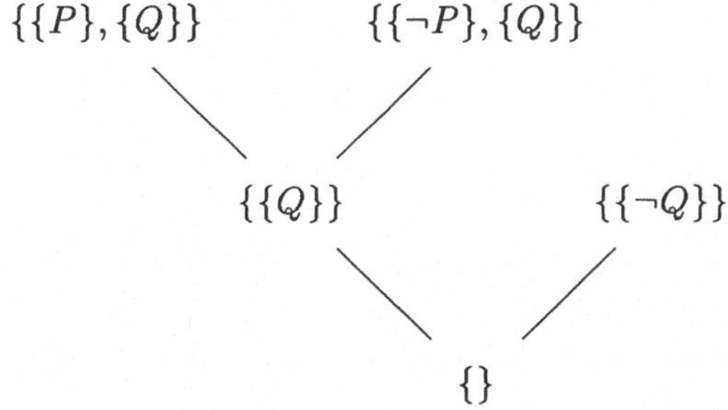

In summary, any consolution calculus must, among its simplifications within the consolution rule, include the elimination of complementary paths. While it is not absolutely necessary, the shortening of paths is understood to be always included in consolution. As an aside we mention that such an inclusion amounts to an extension of the basic calculus. A further extension of consolution might include Prawitz' matrix reduction, but no investigation has been made yet into this possibility.

Note that consolution allows a systematic checking of paths. If we enumerate the paths through a matrix in any given order then a consolution derivation can check them one after the other in this order. A suitable application of shortening of paths will make this systematic checking more efficient by allowing to check more than one path in one step. Systematic checking of paths is not possible with resolution since there at each step the paths are shortened to length 1.

1.4.2 Consolution in first oder logic

The lifting of consolution to first order logic is done in much the same way as it is done for resolution. In this section we give a formal description of consolution for full first order logic.

By a *path set* we mean a finite set of finite sets of literals. We shall use this term even if these sets of literals are not paths in some particular matrix.

Definition 1.4.1
A path set $\mathcal{Q}$ is obtained from a path set $\mathcal{P}$ by *elimination of complementary paths* if there is a set of connections in elements of $\mathcal{P}$ and a most general unifier σ of this set of connections such that $\mathcal{Q}$ is the set of non-complementary elements of $\mathcal{P}\sigma$.

Definition 1.4.2
A path set $\mathcal{Q}$ is obtained from a path set $\mathcal{P}$ by *shortening of paths* if there is a surjective mapping $f : \mathcal{P} \to \mathcal{Q}$ such that $f(p) \subseteq p$ holds for all $p \in \mathcal{P}$.

Definition 1.4.3
A path set $\mathcal{R}$ is obtained from a path set $\mathcal{P}$ by *simplification* if there is a path set $\mathcal{Q}$ such that $\mathcal{Q}$ is obtained from $\mathcal{P}$ by elimination of complementary paths and such that $\mathcal{R}$ is obtained from $\mathcal{Q}$ by shortening of paths.

The inference rule

$$\frac{\mathcal{P} \qquad\qquad \mathcal{Q}}{\mathcal{R}}$$

if there exists a variant $\mathcal{Q}'$ of $\mathcal{Q}$ which does not have any variables in common with $\mathcal{P}$ such that $\mathcal{R}$ is obtained from the product $\mathcal{P}\mathcal{Q}'$ by simplification.

We say then that the path set $\mathcal{R}$ is a *consolvent* of the path sets $\mathcal{P}$ and $\mathcal{Q}$.

Definition 1.4.4
A *derivation* of a matrix M is a finite sequence $(\mathcal{P}_0, \ldots, \mathcal{P}_n)$ of path sets such that the following conditions hold.

1. For all $k = 1, \ldots, n$, the set $\mathcal{P}_k$ equals $\mathcal{P}_c$ for some $c \in M$, or $\mathcal{P}_k$ is a consolvent of $\mathcal{P}_i$ and $\mathcal{P}_j$ for some $i, j < k$.

2. $\mathcal{P}_n = \emptyset$.

We have again a soundness and completeness theorem.

Theorem 1.4.1
A formula in disjunctive normal form is valid if and only if there is a derivation of its matrix by consolution.

Proof: In order to prove correctness, let us assume that F is a formula and M its matrix. Further assume that $(\mathcal{P}_0, \ldots, \mathcal{P}_n)$ is a derivation of M by consolution. To each finite set p of literals we attribute a formula F_p as follows. Let F_p be the disjunction of all literals of p. To each path set $\mathcal{P}$ we attribute a formula $F_{\mathcal{P}}$ as follows. Let $F_{\mathcal{P}}$ be the existential closure of the conjunction of all F_p with $p \in \mathcal{P}$. It then holds that $\{\neg F_{\mathcal{P}}, \neg F_{\mathcal{Q}}\} \models \neg F_{\mathcal{R}}$ if $\mathcal{R}$ is a consolvent of $\mathcal{P}$ and $\mathcal{Q}$ (The proof is exactly the same as the proof for the soundness of the resolution rule). Moreover, $\mathcal{P}_c \models F$ for all $c \in M$. By induction on j it follows that $F_{\mathcal{P}_j} \models F$ for all $j = 0, \ldots, n$. In particular for $j = n$, the set $\mathcal{P}_j$ is the empty set and therefore $F_{\mathcal{P}_j}$ is the empty conjunction, ie., the verum $\top$. So $\top \models F$ which means that F is valid.

For the proof of completeness we shall see in the next section that every set of clauses for which there is a resolution refutation, also has a consolution derivation. From the completeness of resolution it then follows that also consolution is complete.

q.e.d.

In the completeness proof note the duality between proving validity of a formula in disjunctive normal form and proving unsatisfiability of a formula in conjunctive normal form. On the level of clauses and matrices there is no difference between affirmative and refutational proving. In fact any refutational calculus can just as well be formulated in an affirmative way and vice versa, and it would not even make a difference in the codes of implementations.

1.4.3 Simulation of resolution and connection calculus by consolution

If we look at the last proof tree shown in Section 1.4.1 then we see that its path sets contain only one-element paths. If we replace each path with the single literal which it contains then we obtain the tree

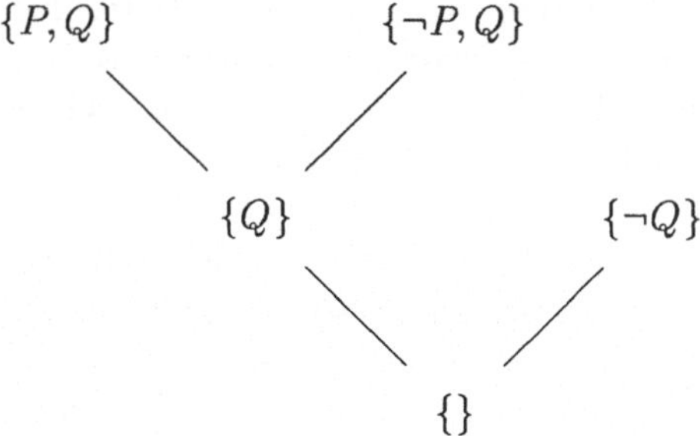

Note that this tree is a resolution refutation of the given clause set. In this way any resolution refutation of a set of clauses can be obtained by consolution. To be more specific, the following holds.

Theorem 1.4.2
Let M be a matrix and let $(c_0, \ldots, c_n)$ be a resolution refutation of M. Then $(\mathcal{P}_{c_0}, \ldots, \mathcal{P}_{c_n})$ is a consolution derivation of M.

Proof: The only non-trivial part of the proof is to show that $\mathcal{P}_e$ is a consolvent of $\mathcal{P}_c$ and $\mathcal{P}_d$ if e is a resolvent of c and d. So let e be a resolvent of c and d and let $c_0 \cup d_0$ be the set of literals resolved upon where $c_0 \subseteq c$ and $d_0 \subseteq d$.[6]. Let d' be the clause obtained from d by seperating apart the variables of c and d. So, d' is a variant of d. Let d'_0 be the corresponding variant of d_0. Then $\mathcal{P}_{d'}$ is a variant of $\mathcal{P}_d$, and $\mathcal{P}_c\mathcal{P}_{d'}$ is the set of all $\{I, J\}$ such that $I \in c$ and $J \in d'$. Let K be the set of all pairs (I, J) such that $I \in c_0$ and $J \in d'_0$. Then the set $\mathcal{Q}$ defined as

$$\{\{I\sigma, J\sigma\} \mid I \in c \text{ and } J \in d' \text{ and not } (I, J) \in K\}$$

is obtained from $\mathcal{P}_c\mathcal{P}_{d'}$ by elimination of complementary paths (choosing as connections the elements of K). For each path $\{I, J\} \in \mathcal{Q}$ we define

$$f(\{I, J\}) := \begin{cases} \{I\} & \text{if } I \notin c_0 \\ \{J\} & \text{otherwise} \end{cases}$$

Then the range of f is $\mathcal{P}_e$. So $\mathcal{P}_e$ is obtained from $\mathcal{Q}$ by shortening of paths. Thus $\mathcal{P}_e$ is a consolvent of $\mathcal{P}_c$ and $\mathcal{P}_d$.

q.e.d.

From this theorem it follows that consolution can simulate resolution step by step (and therefore consolution is complete as indicated in the last section). On the other hand, consolution is more general than resolution because it can handle paths of arbitrary lengths. In fact, resolution can be seen as one consolution strategy where all paths are shortened to length 1 in a certain way at each step. Other strategies of consolution are given by the connection calculi that have been developed so far. They involve paths of arbitrary lengths.

As a further remark we point out that the sets resulting by consolution from paths in a given matrix are not always paths in the matrix as in the example shown above. Rather they may also be the union of such paths even in propositional logic. For illustration, the reader may also think of paths in a matrix that has multiple occurrences of clauses.

Consolution is a method for clause form theorem proving. We have proved that the consolution calculus presented here is complete and sound and that it can simulate resolution step by step. It is more general than resolution because resolution allows only paths of length 1 whereas consolution allows paths of arbitrary length. This advantage becomes most apparent from the fact that consolution allows a systematic checking of paths as any connection calculus does. Such a systematic checking of paths is not possible in resolution since there the paths are shortened to length 1 in every step. On the other hand, consolution provides the powerful tool of the use of lemmata which is present in resolution but lacking in previous connection calculi. A consolvent that has once been derived can be used as a parent in any number of further consolution steps.

Also, it is easy to see that consolution can directly simulate the connection calculus. Namely, in a connection proof, at each stage of the proof there is a current set of paths which still have to be checked for complementarity. Just take

[6]We assume that factorization is included in the resolution rule as in [Rob65]

this path set as the current path set in consolution. An extension step extending the current path p into a clause c in the connection calculus is simulated by building the product $\mathcal{P}\mathcal{P}_c$ of the current path set $\mathcal{P}$ with the set $\mathcal{P}_c$ and then shortening all paths except the extensions of p again. Truncation does not affect the current path set at all. Backward factorization can also be simulated by shortening of paths. Note that by our definition the elements of path sets are sets of literals, not sets of occurrences of literals. Therefore, whenever backward factorization is applicable in the connection calculus, the corresponding path can just be eliminated from the current path set. If we do not allow structure sharing techniques for the encoding in consolution, the encoding will be less efficient than the corresponding encoding using pointers in the connection method. But the loss is only polynomial since in the connection calculus the size of the encoding is never smaller than the number of represented paths.

1.5 The Tableau Calculus TC

In this section we are concerned with a refutation method for full first order predicate logic called the method of *analytic tableaux*. It is a variant of what Evert Beth introduced and called *semantic tableaux* in [Bet55] and [Bet59], and was further developed by Raymond Smullyan in [Smu71]. Here we shall only give a brief description of the method of tableaux and of a tableau calculus which we shall call TC, but we shall not give full proofs of any fundamental properties of the tableau calculus. An introduction to the method of tableaux and the calculus TC and proofs of soundness and completeness of the calculus can be found in the book of Stegmüller and Varga [SVvK84], and, for the case of first order predicate logic without function symbols, also in Smullyan's book [Smu71].

In the tableau method the formulas are categorized according to their syntactical structure. We distinguish five classes of formulas.

1. *Literals* are atomic formulas or negated atomic formulas.

2. *α-formulas* are formulas of any of the following forms. $\neg\neg F$, $F \wedge G$, $\neg(F \vee G)$.

3. *β-formulas* are formulas of any of the following forms. $F \vee G$, $\neg(F \wedge G)$.

4. *γ-formulas* are formulas of any of the following forms. $\forall x F$, $\neg \exists x F$.

5. *δ-formulas* are formulas of any of the following forms. $\exists x F$, $\neg \forall x F$.

Obviously, every formula belongs to exactly one of these five classes. We shall use α as a syntactic variable (meta variable) for α-formulas, β for β-formulas, and so on.

For each α-formula α we define two formulas α_1 and α_2, and for each β-formula β we define two formulas β_1 and β_2. For each γ-formula γ and each variable-free term t we define a formula $\gamma(t)$, and for each δ-formula δ and each constant a we

α	α_1	α_2
$\neg\neg F$	F	F
$F \wedge G$	F	G
$\neg(F \vee G)$	$\neg F$	$\neg G$

β	β_1	β_2
$F \vee G$	F	G
$\neg(F \wedge G)$	$\neg F$	$\neg G$

γ	$\gamma(t)$
$\forall x F$	$F\{x\backslash t\}$
$\neg\exists x F$	$\neg F\{x\backslash t\}$

δ	$\delta(a)$
$\exists x F$	$F\{x\backslash a\}$
$\neg\forall x F$	$\neg F\{x\backslash a\}$

Table 1.1: α-, β-, γ- and δ-formulas

define a formula $\delta(a)$, according to Table 1.1. By $F\{x\backslash t\}$ we denote the result of replacing all free occurrences of the variable x in the formula F by the term t.

For α- and β-formulas it holds that α is true with respect to any given interpretation iff α_1 and α_2 are both true, and β is true iff at least one of the formulas β_1 and β_2 is true. For the problem of satisfiability the following lemma holds.

Lemma 1.5.1
Let S be a satisfiable set of formulas. Then

1. *S does not contain a formula F and its negation $\neg F$.*
2. *If $\alpha \in S$ then $S \cup \{\alpha_1\}$ and $S \cup \{\alpha_2\}$ are satisfiable.*
3. *If $\beta \in S$ then $S \cup \{\beta_1\}$ is satisfiable or $S \cup \{\beta_2\}$ is satisfiable.*
4. *If $\gamma \in S$ then $S \cup \{\gamma(t)\}$ is satisfiable.*
5. *If $\delta \in S$ and a is a constant not occurring in any formula of S then $S \cup \{\delta(a)\}$ is satisfiable.*

Proof: In the cases of α-, β- or γ-formulas we can satisfy the respective set of formulas using the given interpretation satisfying the set S. In the case of a δ-formula the interpretation has to be changed only for the constant a so that a is interpreted as that element of the object domain the existence of which is asserted by saying that the given interpretation satisfies the formula δ. Since a does not occur in any formula of S the set S is still satisfied by this modified interpretation.
q.e.d.

The idea of the method of tableaux is to make repeated use of this lemma in order to derive a contradiction from the assumption that some given set S of formulas is satisfiable.

Definition 1.5.1
A *tableau* for a set S of formulas is a tree T whose nodes are marked with formulas and which obeys the following two conditions.

1. The root of $\mathcal{T}$ is marked with some formula of S.

2. For every node κ of $\mathcal{T}$ which is not a leaf of $\mathcal{T}$ one of the following five cases holds.

   (a) κ has exactly one successor node κ_1. The node κ_1 is marked with some element of S.

   (b) κ has exactly one successor node κ_1. The node κ_1 is marked with α_1 or with α_2 where α occurs somewhere on the path from the root of the tree to the node κ.

   (c) κ has exactly two successor nodes κ_1 and κ_2. The nodes κ_1 and κ_2 are marked with β_1 and β_2, respectively, where β occurs somewhere on the path from the root of the tree to the node κ.

   (d) κ has exactly one successor node κ_1. The node κ_1 is marked with $\gamma(t)$ where γ occurs somewhere on the path from the root of the tree to the node κ, and t is a variable-free term.

   (e) κ has exactly one successor node κ_1. The node κ_1 is marked with $\delta(a)$ where δ occurs somewhere on the path from the root of the tree to the node κ, and a is a constant which neither occurs in any formula of S nor in any formula marking a node on the path from the root of the tree to the node κ.

As an example, in Figure 1.2 a tableau for the set

$$\{\neg((Pa \land Qa) \land Qfa), \forall x Px \land \forall x Qx\}$$

of formulas is given.

We shall call a tableau consisting only of its root marked with some element of a set S of formulas a *trivial tableau* for S. In the tableau calculus TC a tableau is constructed step by step from the root to its leaves. Starting from a trivial tableau for the given set S a sequence of tableaux is constructed by repeatedly appending nodes to leaves of the tableau according to the following rules.

Definition 1.5.2
Let $\mathcal{T}$ be a tableau for a set S of formulas. Let $\mathcal{T}'$ be obtained from $\mathcal{T}$ by appending to some branch B of $\mathcal{T}$ one or two nodes as new successor nodes of the leaf of B. Let these nodes be marked with formulas F_i $(i = 1, \ldots, n)$ where $n = 1$ or $n = 2$.

1. We say that $\mathcal{T}'$ is obtained from $\mathcal{T}$ by *introduction* of an element of S if $n = 1$ and $F_1 \in S$.

2. We say that $\mathcal{T}'$ is obtained from $\mathcal{T}$ by *application of the α-rule* if $n = 1$ and there is an α marking some node of B such that $F_1 = \alpha_1$ or $F_1 = \alpha_2$.

3. We say that $\mathcal{T}'$ is obtained from $\mathcal{T}$ by *application of the β-rule* if $n = 2$ and there is a β marking some node of B such that $F_1 = \alpha_1$ and $F_2 = \beta_2$.

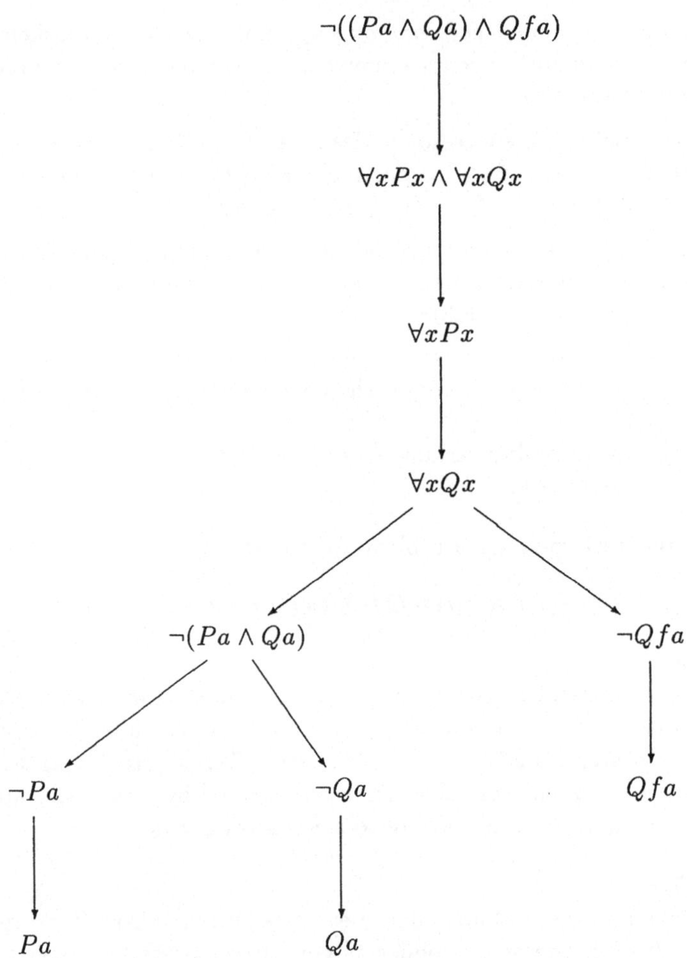

Figure 1.2: A closed tableau for $\{\neg((Pa \wedge Qa) \wedge Qfa), \forall x Px \wedge \forall x Qx\}$

4. We say that $\mathcal{T}'$ is obtained from $\mathcal{T}$ by *application of the γ-rule* if $n = 1$ and there is a γ marking some node of B and a variable-free term t such that $F_1 = \gamma(t)$.

5. We say that $\mathcal{T}'$ is obtained from $\mathcal{T}$ by *application of the δ-rule* if $n = 1$ and the following holds. There is a δ marking some node of B, and there is a constant a which neither occurs in any element of S nor in any formula marking a node on the branch B, such that $F_1 = \delta(a)$.

Every tree obtained from a tableau for S by application of any of these rules is again a tableau for S, and any finite tableau for a set S of formulas can be obtained from a trivial tableau for S by a finite number of applications of these rules.

The α-, β-, γ- and δ-rules are usually written in the following way.

α-RULE $\qquad \dfrac{\alpha}{\alpha_1}$ and $\dfrac{\alpha}{\alpha_2}$

β-RULE $\qquad \dfrac{\beta}{\beta_1 \mid \beta_2}$

γ-RULE $\qquad \dfrac{\gamma}{\gamma(t)}$ where t is a variable-free term

δ-RULE $\qquad \dfrac{\delta}{\delta(a)}$ where a is a new constant

Definition 1.5.3
A branch of a tableau is said to be *closed* iff it contains a formula F and the negation of F. A tableau is said to be *closed* iff each of its branches is closed.

Definition 1.5.4
By a *refutation* of a set S of formulas in the tableau calculus TC we mean a finite closed tableau for S.

The tableau depicted in Figure 1.2 is an example of a refutation.

If a set S of formulas is satisfiable and $\mathcal{T}$ is a tableau for S then at least one branch of $\mathcal{T}$ is satisfiable, i.e., the set of formulas marking its nodes is satisfiable. In the case of finite tableaux this can be proved from Lemma 1.5.1 by induction on the number of applications of rules needed to obtain the tableau $\mathcal{T}$ from a trivial tableau for S. Since every closed branch is unsatisfiable every set of formulas having a refutation in the tableau calculus is unsatisfiable (soundness of the tableau calculus). In fact, the tableau calculus is also complete.

Proposition 1.5.1
For a set S there is a refutation in the tableau calculus iff S is unsatisfiable.

The proof of this theorem of adequacy of the tableau calculus can be found in the literature, e.g., in Smullyan [Smu71] and Stegmüller and Varga [SVvK84].

Intuitively, the method of tableaux can be viewed as assuming the satisfiability of the given set S of formulas and then repeatedly applying Lemma 1.5.1 thus getting a nested case analysis until we prove a contradiction in each of the cases. Each branch of the tree corresponds to a case in the case analysis. For each choice of a β-formula there is a branching point of the case analysis, corresponding to a branching point in the tree. Each closed branch in the tree corresponds to a contradiction in the case analysis.

1.6 The Sequent Calculus

In this section we present Gerhard Gentzen's sequent calculus which was introduced in [Gen35]. Since Gentzen considers only pure first order logic without function symbols in [Gen35] we shall have to adapt his calculus slightly to incorporate function symbols. The only changes which are necessary are in the rules $\forall$-IA and $\exists$-IS where we have to allow arbitrary terms t instead of his free object variables.

The sequent calculus does not work directly with formulas but with so-called sequents. A *sequent* is a string of characters of the form

$$A_1, \ldots, A_\mu \Longrightarrow B_1, \ldots, B_\nu$$

where $\mu, \nu \geq 0$ and $A_1, \ldots, A_\mu, B_1, \ldots, B_\nu$ are formulas. The formulas $A_1, \ldots, A_\mu$ form the *antecedent*, and the formulas $B_1, \ldots, B_\nu$, the *succedent* of the sequent. The sequent has the same informal meaning as the formula

$$A_1 \wedge \cdots \wedge A_\mu \rightarrow B_1 \vee \cdots \vee B_\nu.$$

We use capital Latin letters as syntactic variables to denote formulas and capital Greek letters as syntactic variables to denote finite sequences $A_1, \ldots, A_\mu$ of formulas separated by commata. So, every sequent has the form $\Gamma \Longrightarrow \Theta$. Moreover, for convenience we assume that the set of variables is partitioned in two denumerably infinite subsets, the set of *free variables* and the set of *bound variables*. We shall use a, b, c as syntactic variables to denote free variables and x, y, z to denote bound variables. Now, the sequent calculus has as axioms the sequents consisting of one formula in the antecedent and the same formula in the succedent. In addition there are two kinds of inference rules, structural and operational inference rules. For each propositional connective and each quantifier, there is one operational rule for its introduction in the antecedent and one operational rule for its introduction in the succedent:

There is an additional restriction for the applications of the rules $\forall$-IS and $\exists$-IA. Namely, the *eigenvariable* a must not occur in the conclusion, i.e., neither

Table 1.2: Gentzen's sequent calculus

AXIOMS $\quad D \Longrightarrow D$

Structural inferences:

THINNING
$$\frac{\Gamma \Longrightarrow \Theta}{D, \Gamma \Longrightarrow \Theta} \qquad \frac{\Gamma \Longrightarrow \Theta}{\Gamma \Longrightarrow \Theta, D}$$

CONTRACTION
$$\frac{D, D, \Gamma \Longrightarrow \Theta}{D, \Gamma \Longrightarrow \Theta} \qquad \frac{\Gamma \Longrightarrow \Theta, D, D}{\Gamma \Longrightarrow \Theta, D}$$

INTERCHANGE
$$\frac{\Delta, D, E, \Gamma \Longrightarrow \Theta}{\Delta, E, D, \Gamma \Longrightarrow \Theta} \qquad \frac{\Gamma \Longrightarrow \Theta, E, D, \Lambda}{\Gamma \Longrightarrow \Theta, D, E, \Lambda}$$

CUT
$$\frac{\Gamma \Longrightarrow \Theta, D \qquad D, \Delta \Longrightarrow \Lambda}{\Gamma, \Delta \Longrightarrow \Theta, \Lambda}$$

Operational inferences:

$\neg$-Introductions: $\neg$-IA $\quad$ $\neg$-IS
$$\frac{\Gamma \Longrightarrow \Theta, A}{\neg A, \Gamma \Longrightarrow \Theta} \qquad \frac{A, \Gamma \Longrightarrow \Theta}{\Gamma \Longrightarrow \Theta, \neg A}$$

$\wedge$-Introductions: $\wedge$-IA $\quad$ $\wedge$-IS
$$\frac{A, \Gamma \Longrightarrow \Theta}{A \wedge B, \Gamma \Longrightarrow \Theta} \quad \frac{B, \Gamma \Longrightarrow \Theta}{A \wedge B, \Gamma \Longrightarrow \Theta} \qquad \frac{\Gamma \Longrightarrow \Theta, A \qquad \Gamma \Longrightarrow \Theta, B}{\Gamma \Longrightarrow \Theta, A \wedge B}$$

$\vee$-Introductions: $\vee$-IA $\quad$ $\vee$-IS
$$\frac{A, \Gamma \Longrightarrow \Theta \qquad B, \Gamma \Longrightarrow \Theta}{A \vee B, \Gamma \Longrightarrow \Theta} \qquad \frac{\Gamma \Longrightarrow \Theta, A}{\Gamma \Longrightarrow \Theta, A \vee B} \quad \frac{\Gamma \Longrightarrow \Theta, B}{\Gamma \Longrightarrow \Theta, A \vee B}$$

$\rightarrow$-Introductions: $\rightarrow$-IA $\quad$ $\rightarrow$-IS
$$\frac{\Gamma \Longrightarrow \Theta, A \qquad B, \Delta \Longrightarrow \Lambda}{A \rightarrow B, \Gamma, \Delta \Longrightarrow \Theta, \Lambda} \qquad \frac{A, \Gamma \Longrightarrow \Theta, B}{\Gamma \Longrightarrow \Theta, A \rightarrow B}$$

$\forall$-Introductions: $\forall$-IA $\quad$ $\forall$-IS
$$\frac{F\{x \backslash t\}, \Gamma \Longrightarrow \Theta}{\forall x F, \Gamma \Longrightarrow \Theta} \qquad \frac{\Gamma \Longrightarrow \Theta, F\{x \backslash a\}}{\Gamma \Longrightarrow \Theta, \forall x F}$$

with eigenvariable condition

in F nor in any formula of Γ or Θ. Moreover, there is the restriction that no free variable occurs bound and no bound variable occurs free in any formula of any of the occurring sequents and t does not contain any bound variables.

For each application of a rule, the sequents above the horizontal bar are called *premises* and the sequent below the horizontal bar is called the *conclusion*. The formula D in the cut rule is called the *cut formula*. A *derivation* in the sequent calculus is a finite sequence $(\mathcal{S}_1, \ldots, \mathcal{S}_n)$ of sequents such that each sequent $\mathcal{S}_k$ of the sequence is an axiom or the conclusion of an application of a rule whose premises are in $\{\mathcal{S}_1, \ldots, \mathcal{S}_{k-1}\}$. The last sequent of the derivation is also called the *endsequent* of the derivation. We pose the further restriction on derivations that all sequents except the endsequent occur as premises in the derivation. A *derivation of a sequent* is a derivation which has this sequent as its endsequent, and a *derivation of a formula* F is a derivation of the sequent $\Longrightarrow F$.

Gentzen has proved in [Gen35] that the sequent calculus is sound and complete, by showing its equivalence to a Frege-Hilbert calculus. Furthermore, the paper contains his famous *Hauptsatz* or *cut elimination theorem* stating that every deduction of a sequent $\mathcal{S}$ in the sequent calculus can be transformed to a deduction of $\mathcal{S}$ in which no application of the cut rule occurs. So, the sequent calculus is complete also without the cut rule.

In mechanical theorem proving, the sequent calculus seems to be more adequate for interactive theorem proving than for automatic theorem proving because it is more difficult to build unification into the sequent calculus than it is for resolution. D. Miller and A. Felty [MF86] have implemented a system using the sequent calculus without the cut rule. Their system allows easy conversion of a resolution refutation to a proof in the sequent calculus.

1.7 Natural Deduction

The calculus of natural deduction was introduced by Gerhard Gentzen in [Gen35]. The motivation Gentzen had when he introduced this calculus was to formalize and model the way mathematicians make proofs of mathematical theorems. Most calculi introduced up to that time were Frege-Hilbert calculi. In a Frege-Hilbert calculus, a derivation starts from valid formulas as axioms, and from these axioms new valid formulas are derived according to given rules. In contrast to this, in a natural deduction calculus, the derivation starts from assumptions from which new formulas are derived that depend on these assumptions. There are rules for making a derived formula independent of assumptions that have been made.

In the field of automated theorem proving, natural deduction calculi have been used by several authors to implement their systems. Most of these systems are not fully automatized. Rather, they are either proof checking systems, or interactive theorem provers with automatic components to fill in the gaps between the proof steps suggested by the user. Such systems are, for example, Nicolas de Bruijn's AUTOMATH [dB80], and the NuPRL system of Robert Constable et. al. [C$^+$86]. Some of these systems draw a connection between natural deduction and variants or extensions of Martin-Löf's type theory. Natural deduction style presentations of proofs have be used to formulate a framework for a wide class of logics at the University of Edinburgh (see [Bur87], [AHM87], [HHP87]). See also the survey of mechanical support for formal reasoning by P. A. Lindsay, [Lin88].

The calculus we shall consider here differs from the calculus presented in [Gen35] in three ways. First, it is here adapted to predicate logic with function symbols whereas Gentzen considered only predicate logic without function symbols. Second, Gentzen introduces derivations as trees. In his tree representation, the assumptions that a formula in the derivation depends upon are not attached as labels to the same node of the tree to which the formula itself is attached. Rather, Gentzen uses a structure sharing technique. Since we are interested not only in tree derivations but also in derivations of arbitrary dag form, it seemed to be more convenient to define a derivation as a finite sequence of sequents where the succedent of each sequent consists of only one formula, and the antecedent consists of all the assumptions on which this formula depends. So, every element of the sequence is in a sense self-contained needing no external information to determine its meaning. The loss in efficiency caused by rejecting structure sharing techniques is only polynomial and irrelevant for p-simulatability.

We use as propositional connectives the nullary connective $\bot$ (falsum), the unary connective $\neg$, and the binary connectives $\wedge$, $\vee$, and $\rightarrow$. The quantifiers are $\forall$ and $\exists$. We have, for each logical symbol, an introduction rule and an elimination rule. The axioms and rules of the calculus are the following.

Definition 1.7.1
Let R be a rule of the calculus of natural deduction. Let R_1 be the result of consistently replacing in R all capital letters with formulas, the symbol 'x', with a bound variable x, the symbol 'a' with a free variable a, and the symbol 't' with a term t not containing any bound variables. Finally,[7] let R_2 be obtained from R_1 by replacing $F\{x\backslash a\}$ with the result of replacing x with a in F in the case of $\forall$-I or $\exists$-E, and by replacing $F\{x\backslash t\}$ with the result of replacing x with t in F in the case of $\forall$-E or $\exists$-I. Otherwise let $R_2 \stackrel{\text{def}}{=} R_1$. Then we say that R_2 is an *instance* of R.

Each instance of a rule of the calculus of natural deduction consists of one, two, or three premises arranged horizontally next to each other, together with a horizontal line below these premises and the conclusion below the horizontal line. Each premise is a vertical column of one or two formulas where the upper formula is bracketed if there are two formulas. The conclusion is a formula.

We shall now state what we mean by a derivation in the calculus of natural deduction. Intuitively, the assertions derived in the course of a derivation are of the form 'under certain assumptions a certain formula H is true'. They are called 'judgements'. For each such judgement, the set S of assumptions is a finite set of formulas. Formally, we define a *judgement* to be a sequent $\Gamma \Longrightarrow H$ where Γ

[7]Here, let F be the formula by which the symbol 'F' was replaced when passing from R to R_1.

Table 1.3: Gentzen's system of natural deduction

Axioms:

ASSUMPTION	$A \Longrightarrow A$
TERTIUM NON DATUR	$\Longrightarrow A \vee \neg A$

Rules:

EX FALSO QUODLIBET $\dfrac{\bot}{D}$

$\neg$-I	$\dfrac{\begin{matrix}[A]\\ \bot\end{matrix}}{\neg A}$	$\neg$-E	$\dfrac{A \qquad \neg A}{\bot}$
$\wedge$-I	$\dfrac{A \qquad B}{A \wedge B}$	$\wedge$-E	$\dfrac{A \wedge B}{A} \qquad \dfrac{A \wedge B}{B}$
$\vee$-I	$\dfrac{A}{A \vee B} \qquad \dfrac{B}{A \vee B}$	$\vee$-E	$\dfrac{A \vee B \qquad \begin{matrix}[A]\\ C\end{matrix} \qquad \begin{matrix}[B]\\ C\end{matrix}}{C}$
$\rightarrow$-I	$\dfrac{\begin{matrix}[A]\\ B\end{matrix}}{A \rightarrow B}$	$\rightarrow$-E	$\dfrac{A \qquad A \rightarrow B}{B}$
$\forall$-I	$\dfrac{F\{x\backslash a\}}{\forall x F}$ with eigenvariable condition	$\forall$-E	$\dfrac{\forall x F}{F\{x\backslash t\}}$
$\exists$-I	$\dfrac{F\{x\backslash t\}}{\exists x F}$	$\exists$-E	$\dfrac{\exists x F \qquad \begin{matrix}[F\{x\backslash a\}]\\ C\end{matrix}}{C}$ with eigenvariable condition

consists of a finite number of formulas separated by commata, and H is a formula. The members of Γ are the assumptions of the judgement. The informal meaning of such a judgement is that H is true under the assumption that all formulas of Γ are true. Now, let us use the following notation. If T is the empty set and H is a formula then $\overset{[T]}{H}$ is defined to be the formula H. If T is a singleton set of formulas, $T = \{A\}$, and H is a formula, then $\overset{[T]}{H}$ is defined to be the figure $\overset{[A]}{H}$. Then each instance R of a rule of the calculus of natural deduction has the form $\overset{[T]}{H}$ where T is an (empty or singleton) set of formulas and H is a formula, or R has the form

$$\frac{\overset{[T_1]}{H_1} \qquad \cdots \qquad \overset{[T_n]}{H_n}}{H}$$

where $H_1, \ldots, H_n$ and H are formulas and $T_1, \ldots, T_n$ are (empty or singleton) sets of formulas. If Γ is a finite sequence of formulas separated by commata, then we denote by $\tilde{\Gamma}$ the set of all members of Γ.

Definition 1.7.2
Let $\Gamma_1, \ldots, \Gamma_n$, and Γ be finite sets of formulas, and let $H_1, \ldots, H_n$, and H be formulas. Let $\mathcal{J}_1 \stackrel{\text{def}}{=} \Gamma_1 \Longrightarrow H_1, \ldots, \mathcal{J}_n \stackrel{\text{def}}{=} \Gamma_n \Longrightarrow H_n$, and $\mathcal{J} \stackrel{\text{def}}{=} \Gamma \Longrightarrow H$. Let

$$\frac{\overset{[T_1]}{H_1} \qquad \cdots \qquad \overset{[T_n]}{H_n}}{H}$$

be an instance of a rule R of the natural deduction calculus. Further assume that $\tilde{\Gamma} = \bigcup_{j=1}^{n}(\tilde{\Gamma}_j \setminus T_j)$. In the case of a $\forall$-I assume that the *eigenvariable* a does not occur in Γ or in H. In the case of an $\exists$-E assume that the *eigenvariable* a does not occur in Γ or in H or in H_1. Then we say that the judgement $\mathcal{J}$ is obtained from the judgements $\mathcal{J}_1, \ldots, \mathcal{J}_n$ by *application of the rule R*.

Definition 1.7.3
A *derivation* of a formula F in the calculus of natural deduction is a finite sequence $(\mathcal{J}_1, \ldots, \mathcal{J}_r)$ of judgements such that each $\mathcal{J}_j$ is either an axiom or obtained from judgements in $\{\mathcal{J}_1, \ldots, \mathcal{J}_{j-1}\}$ by an application of a rule of the natural deduction calculus, and such that $\mathcal{J}_r$ is the judgement $\Longrightarrow F$.

The natural deduction calculus is sound and complete for first order predicate logic as has been shown by Gentzen in [Gen35] (for predicate logic without function symbols).

Proposition 1.7.1
A formula F is valid iff there exists a derivation of F in the natural deduction calculus.

Table 1.4: A Frege–Hilbert calculus

The Axiom Schemes:

A1	$A \vee \neg A$
A2	$A \rightarrow A$
A3	$A \rightarrow (B \rightarrow A)$
A4	$(A \rightarrow (A \rightarrow B)) \rightarrow (A \rightarrow B)$
A5	$(A \rightarrow (B \rightarrow C)) \rightarrow (B \rightarrow (A \rightarrow C))$
A6	$(A \rightarrow B) \rightarrow ((B \rightarrow C) \rightarrow (A \rightarrow C))$
A7	$(A \wedge B) \rightarrow A$
A8	$(A \wedge B) \rightarrow B$
A9	$(A \rightarrow B) \rightarrow ((A \rightarrow C) \rightarrow (A \rightarrow (B \wedge C)))$
A10	$A \rightarrow (A \vee B)$
A11	$B \rightarrow (A \vee B)$
A12	$(A \rightarrow C) \rightarrow ((B \rightarrow C) \rightarrow ((A \vee B) \rightarrow C))$
A13	$(A \rightarrow B) \rightarrow ((A \rightarrow \neg B) \rightarrow \neg A)$
A14	$\neg A \rightarrow (A \rightarrow B)$
A15	$\forall x F \rightarrow F\{x \backslash t\}$
A16	$F\{x \backslash t\} \rightarrow \exists x F$

The Rules:

R1	$\dfrac{A \quad A \rightarrow B}{B}$	
R2	$\dfrac{A \rightarrow F\{x \backslash a\}}{A \rightarrow \forall x F}$	with eigenvariable condition
R3	$\dfrac{F\{x \backslash a\} \rightarrow A}{\exists x F \rightarrow A}$	with eigenvariable condition

1.8 A Frege-Hilbert Calculus

In this section we describe a Frege-Hilbert calculus that has been used by Gerhard Gentzen to prove the equivalence of his natural deduction calculus with Frege-Hilbert calculi. A derivation of a formula in a Frege-Hilbert calculus is a sequence of valid formulas. Starting from axioms, new valid formulas are derived according to given rules. The axioms and rules in the calculus we consider here, are the following.

There is the restriction on the critical rules R2 and R3 that the *eigenvariable* a must not occur in the conclusion.

Definition 1.8.1
A *derivation* of a formula F is a finite sequence $(F_1, \ldots, F_n)$ of formulas such that $F_n \equiv F$, and each F_k is an axiom or the conclusion of an instance of a rule whose premises are in $\{F_1, \ldots, F_{k-1}\}$. In the case of the rules R2 and R3, the

eigenvariable a must not occur in the conclusion.

The following soundness and completeness theorem holds for the Frege-Hilbert calculus.

Proposition 1.8.1
Let F be a closed formula. Then there is a derivation of F in the Frege-Hilbert calculus if and only if F is valid.

Chapter 2

Comparison of Calculi for First Order Logic

2.1 Known Results on the Complexity of Calculi

From Chapter 1, we are now familiar with a couple of calculi which have a few common features. Namely, there is a notion of a *derivation* or *proof* (or *refutation*) of a formula. A derivation $\mathcal{D}$ is a string over some fixed alphabet, or some object such as a tree that can easily be transformed to such a string representation. Now, the notion of a derivation always has the property that, for a given formula F and a given string $\mathcal{D}$, it is decidable whether $\mathcal{D}$ is a derivation of F. And, in fact, in all calculi that we consider here, this is decidable in a time polynomial in the number of symbol occurrences of $\mathcal{D}$. Moreover, a soundness- and completeness-theorem holds stating that a formula is valid (resp., unsatisfiable, in the case of refutation calculi) if and only if it has a derivation in the calculus.

Results in First Order Logic

There has been an interest for a long time in the question whether the existence of a derivation for a given formula is decidable. For first order predicate logic it has been shown (Gödel [Göd31], Church [Chu36]) that this problem is not decidable. In fact, the set of valid formulas is recursively enumerable, and it is complete in the set of recursively enumerable sets in the following sense. For every recursively enumerable set S over some alphabet A there is a recursive function ϕ from A^* to the set of formulas such that a string s is in S iff $\phi(s)$ is valid (see Gödel [Göd31], Theorem X). In the same sense, the set of valid formulas is complete in the set of Σ_1-sets[1] in the arithmetic hierarchy since the Σ_1-sets are exactly the recursively

[1] A Σ_1-*formula* is a formula of first order predicate logic in prenex normal form that contains no predicate symbols other than $=$, and no function symbols other than $+$, $\cdot$ and 1, and no universal quantifier $\forall$. If x is a variable, and F is a Σ_1-formula in which no variable other than x occurs free then let S_F be the set of natural numbers n such that $F\{x\backslash n\}$ is true in the standard

enumerable sets.[2]

In [Gen35], Gerhard Gentzen described in detail how his calculus of natural deduction and his sequent calculus can be simulated by Frege-Hilbert calculi and vice versa. We shall see in Section 3.7 that his simulations can be done at polynomial cost. His Hauptsatz or cut elimination theorem says that any derivation of a formula in the sequent calculus can be transformed to a derivation of the same formula in the sequent calculus without the cut rule. However, this simulation is not polynomial. But the size of the cut-free derivation is less than $f(n)$ where n is the size of the given derivation and $f(0) \stackrel{\text{def}}{=} 1$ and $f(k+1) \stackrel{\text{def}}{=} 2^{f(k)}$ (see M. M. Richter [Ric78]). A much more concise proof of the cut elimination theorem for a similar logical calculus, the calculus of positive and negative parts, can be found in Kurt Schütte's book Proof Theory [Sch77].

Results in Propositional Logic

In the case of propositional logic, the validity of a formula is decidable. The simplest decision procedure is the generation of a table of truth values for the formula which takes time exponential w.r.t. the length of the formula. In stating some of the known results on the lengths of shortest proofs of a formula in calculi of propositional logic, we shall make use of the terminology of Stephen Cook and Robert Reckhow in [CR74].

Definition 2.1.1
A *proof system* is a surjective in polynomial time computable function F from the set of strings over some finite alphabet to the set of valid formulas. Every string w is said to be a *proof* of the formula $F(w)$.

Every formal calculus can be considered as a proof system.

Definition 2.1.2
A proof system is said to be *super* if there is a polynomial p such that the length of the shortest proof of any valid formula is less than $p(n)$ where n is the length of the formula. A proof system S_1 can *p-simulate* a proof system S_2 if there is a polynomial p such that for every natural number n and for every formula F the following holds. If there is a proof of F in S_2 whose length is n then there is a proof of F (or of a suitable translation[3] of F) in S_1 whose length is less than $p(n)$.

---

interpretation of F in the domain of natural numbers. Then a set S of natural numbers is said to be a Σ_1*-set* if there is a Σ_1-formula F such that $S = S_F$.

[2] This is the undecidability theorem for Hilbert's 10th problem. The theorem by M. Davis, H. Putnam, J. Robinson, Ju. Matijasevič and G. Čudnovskij from 1970 states that a set of natural numbers is a Σ_1-set iff it is recursively enumerable. A proof can be found in [Man77], Chapter VI.

[3] A translation of F from one language to another is necessary if the sets of propositional connectives in S_1 and S_2 do not coincide or if one of the two systems allows arbitrary propositional formulas and the other is restricted to normal form.

Some of the first results on the complexity of derivation in calculi of propositional logic were presented by G. S. Tseitin in 1966 (Engl. transl. of his paper [Tse70]). He considers resolution and introduces an additional rule, the *extension rule*. An extension step consists in adding three clauses[4] $\{J, K\}$, $\{J, L\}$ and $\{\bar{J}, \bar{K}, \bar{L}\}$ to the set of clauses where J is a literal that does not already occur in the set of clauses. He considers the length of a derivation as well as the number of occurrences of clauses in the derivation tree. The latter equals the number of occurrences of clauses in a derivation that is subject to the restriction that every occurrence of a clause be used at most once as a parent clause. Let us call resolution with this restriction *tree resolution*. A resolution derivation is *regular* if there is no literal that is resolved upon twice in the same branch of the derivation tree. Then Tseitin's main results are the following.[5]

1. Tree resolution cannot p-simulate regular resolution.
2. Tree resolution cannot p-simulate tree resolution with extension.
3. Regular resolution cannot p-simulate regular resolution with extension.
4. Tree resolution and regular resolution are not super.

Furthermore, Tseitin indicates in that paper how resolution with extension can p-simulate Gentzen's sequent calculus (with the cut rule), and how resolution (without extension) can p-simulate Gentzen's sequent calculus without the cut rule. Since the sequent calculus allows arbitrary propositional formulas, Tseitin has to transform them to clausal form first in order to make them treatable by resolution. We shall see in Section 2.2 how this transformation is done. In Chapter 3 we shall generalize Tseitin's idea to first order predicate logic.

Stephen A. Cook has proved in [Coo71] that the set of valid formulas is co-NP-complete[6], i.e., its complement is NP-complete. So there is a super proof system iff $NP = coNP$, and iff the set of tautologies is in NP. The question whether a super proof system exists is still open today. Moreover, Cook proved that the set of valid formulas in disjunctive normal form is co-NP-complete, and also the set of valid formulas in disjunctive normal form with clauses of cardinalities ≤ 3. His proof method consists in using formulas of propositional logic to encode the working of a non-deterministic time- and space-bounded Turing machine. Today, his results are known under the names 'NP-completeness of SAT' and 'NP-completeness of 3-SAT'.

In [CR74][7], Stephen Cook and Robert Reckhow made a comparison of several proof calculi for propositional logic. A detailed comparison of proof systems, with proofs of the new results, was given by Robert Reckhow in his PhD thesis [Rec76]. Roughly speaking, Cook's and Reckhow's results are the following. Frege systems,

[4]If L is a literal then $\bar{L}$ denotes the literal which is complementary to L.

[5]Tseitin actually gives explicit estimates of the lengths of proofs.

[6]This is what he actually has proved, although the result he states in his paper is slightly weaker.

[7]Pay attention to the corrections in Sigact News.

natural deduction systems, and sequent calculi are equivalent in the sense that they can p-simulate each other. Moreover they can p-simulate resolution, but can be p-simulated by resolution with extension. On the other hand, they show that the tableau calculus, the Davis-Putnam procedure, semantic trees and tree resolution are not super. The latter two can p-simulate the tableau calculus, but not vice versa. There are a lot more results which can be found in their papers [CR74], [Rec76], and [CR79]. Since we are mainly interested in first order logic here, we shall not go into any more details concerning known results for propositional logic. Rather, we refer the reader to the cited literature.

The most remarkable advance since has been the proof by Armin Haken [Hak85] that resolution is not a super proof system. He considered a sequence of unsatisfiable formulas, called the pigeonhole formulas, and proved that the shortest resolution refutations for them have a complexity exponential in the cubic root of the length of the formula. For the pigeonhole formulas, Stephen Cook had proved in [Coo76] that they have polynomial size refutations in resolution with extension. So, from Haken's result it follows that resolution cannot p-simulate resolution with extension. Alasdair Urquhart [Urq87] and Samuel Buss [Bus87] proved that resolution cannot p-simulate Frege systems. Frege systems allow polynomial size proofs of the pigeonhole formulas.

Wolfgang Bibel [Bib89] gave polynomial size proofs of the pigeonhole formulas in the connection method supplemented with the feature of Prawitz' matrix reduction and the feature of renaming of propositional variables. The latter says that, from a formula F that has been proved to be valid, we can infer any formula G that is obtained from F by renaming its propositional variables. Murray and Rosenthal [MR89] have given refutations of polynomial size of the pigeonhole formulas by path dissolution augmented by variable renaming. Path dissolution is very similar to the connection method and also to the Davis-Putnam procedure and to Prawitz' matrix reduction.

A Remark

Note that renaming of propositional variables is not a sound rule in propositional logic, i.e., $F \vdash G$ does not imply $F \models G$. But it preserves validity, and this is all that is needed. Note the similarity to the substitution rule for Frege systems. Renaming can be viewed as a special case of the substitution rule. It would be interesting to know whether the use of non-sound rules has the potential of adding a strength to proof systems that can otherwise not be achieved. More precisely, assume we are given some fixed logic. Let $\mathcal{F}$ be the set of its formulas and $\mathcal{F}^*$ the set of finite sequences of formulas. Let us denote by the term *polynomial deduction rule* any set $R \subseteq \mathcal{F}^* \times \mathcal{F}$ which is decidable in polynomial time. A polynomial deduction rule R is *sound* if $S \models F$ for all $(S, F) \in R$. A *derivation* of a formula F via a polynomial deduction rule R is a finite sequence $\mathcal{D}$ of formulas ending in F such that, for each formula G in $\mathcal{D}$, there is an $(S, G) \in R$ such that all formulas of S are elements of the sequence $\mathcal{D}$ and precede G in $\mathcal{D}$. If R is a polynomial deduction rule then the *proof system induced* by R is the proof system mapping every derivation of a formula F via R to F. Is there a proof system induced by a polynomial deduction rule that cannot be p-simulated by any proof system induced by a sound polynomial deduction rule? Note that renaming of propositional variables in propositional logic — and, more generally, the substitution rule — can be directly p-simulated by a sound rule in second order logic by replacing each formula with its universal closure w.r.t. the propositional variables.

2.2 Transformation to Clausal Form

It has already been mentioned that every formula F of first order predicate logic can be transformed to a formula F' in conjunctive normal form, called its conjunctive Skolem normal form, such that F is satisfiable iff F' is satisfiable (see Thoralf Skolem [Sko20]; Engl. transl. in [Hei67], pp. 252–263).[8] Most theorem proving systems implemented so far use the transformation that can be found in most logic textbooks. It has the disadvantage that it has an exponential worst-case behaviour in terms of space as well as time resources. Moreover, it disrupts the natural structure of the formula. In this section we briefly describe a transformation to clausal form that does not have these disadvantages.

Let us first have a look at the usual transformation of a formula F to its conjunctive Skolem normal form. The idea of the transformation is the following. First use the rules for interchanging quantifiers and propositional connectives to obtain a prenex normal form of F, i.e., a formula consisting of a number of quantifier prefixes followed by a formula that does not contain any quantifiers. This prenex normal form is semantically equivalent to F. The next step is to move all negation signs inside using de Morgan's rules and then deleting double negation signs. The result is a formula consisting of quantifier prefixes followed by a formula built from literals by repeated application of disjunction and conjunction. In a next step, repeated application of the distributive law for $\vee$ and $\wedge$ gives a formula consisting of quantifier prefixes followed by some (arbitrarily nested) conjunction H of (arbitrarily nested) disjunctions of literals. The obtained formula is still semantically equivalent to F. Finally, every existential quantifier $\exists x$ is deleted, and every occurrence of x is replaced in H by a term $f(x_1, \ldots, x_k)$ where f is a new function symbol (not already occurring in H) and $x_1, \ldots, x_k$ are the variables bound by the $\forall$-quantifiers preceding $\exists x$ in the formula. This step is called skolemization. The resulting formula is the universal closure of a conjunction of disjunctions of literals to which we can apply the associativity law for $\vee$ and $\wedge$. It is not semantically equivalent to F any more, but it is satisfiable iff F is satisfiable. Finally, interchanging occurrences of $\wedge$ and $\forall$ yields the required formula in clausal form.

Note that there is only one class of transformation steps in this transforma-

[8] Skolem actually considers only pure predicate logic without function symbols, and he transforms an arbitrary formula of that logic to a formula of the prefix-class $\forall^*\exists^*$ (Theorem 1 in [Sko20]). He then introduces what is known today as Skolem functions purely on the meta-level without any counterpart in the formal system (see the proof of Theorem 2 in [Sko20]).

tion which increases the number of occurrences of atomic formulas in the whole formula. This is the application of the distributive law. The skolemization does not increase the number of atomic formulas, but it may increase the length of each atomic formula by a factor which is at most equal to the number of occurrences of quantifiers in the given formula F. So the increase of the length of the formula due to skolemization is at most quadratic. On the other hand, the increase of the length of a formula due to applications of the distributive law may be as bad as exponential as the following example shows. Let the formula F of propositional logic be given by

$$F \equiv (A_1 \wedge B_1) \vee \cdots \vee (A_n \wedge B_n).$$

It can be proved that there is no formula in conjunctive normal form that is semantically equivalent to F and that contains less than 2^n clause formulas as conjuncts. So any repeated application of the distributive law or of any other rule preserving semantic equivalence in order to obtain a clausal form of F will necessarily lead to an exponential explosion of the length of F.

However, there is another transformation to clausal form that does not suffer from this exponential explosion. In fact, the length of the formula increases at most by a factor given by the maximal depth of nesting of quantifiers in the formula. The number of atomic subformulas even increases only by a constant factor. The idea of this transformation to *definitional form* is to introduce, for each subformula G of F, a new predicate symbol P_G whose arity equals the number of free variables of G. Let L_G be the formula $P_G(x_1, \ldots, x_k)$ where $x_1, \ldots, x_k$ are the free variables of G. For each subformula G of F we introduce a formula D_G, intuitively serving as a definition defining L_G as an abbreviation for G, as follows. If G is atomic then let D_G be the universal closure of $L_G \leftrightarrow G$. If G is non-atomic, $G \equiv \circ(G_1, \ldots, G_n)$, then

$$D_G \stackrel{\text{def}}{=} \forall x_1 \ldots \forall x_k (L_G \leftrightarrow \circ(L_{G_1}, \ldots, L_{G_n})).$$

Here $\circ$ denotes either a propositional connective or a quantifier prefix. Let D be the conjunction of all D_G taken over all subformulas G of F.

Then the formula F is valid iff $D \rightarrow L_F$ is valid, and F is satisfiable iff $D \wedge L_F$ is satisfiable. Moreover, $D \models L_F \leftrightarrow F$. Now, the point is that $D \rightarrow L_F$ or $D \wedge L_F$, respectively, is a formula which is already very close to clausal form. Its transformation to clausal form using the standard procedure described above leads only to a moderate increase of its length. The total space as well as time requirements for the whole transformation of F are at most of the order of the product of the length of F and the maximal depth of nesting of quantifiers in F. Also, the structure of the formula is not disrupted as by the usual transformation to clausal form. Therefore the transformation to definitional form and similar transformations to normal form are also called 'structure-preserving' in the literature.

To be more precise, let us assume that we have a language $\mathcal{L}$ of first order logic with function symbols, and that the variables in $\mathcal{L}$ are enumerated in some fixed order. Now, for each formula G of $\mathcal{L}$, we introduce a new k-ary predicate symbol P_G where k is the number of variables having free occurrences in G. We require that P_G is not a symbol of $\mathcal{L}$ and that P_G and P_H are distinct for distinct formulas

G and H. Similarly, for each formula G of $\mathcal{L}$ which starts with a quantifier, we introduce a new k-ary function symbol f_G where k is the number of variables having free occurrences in G. The function symbol f_G will serve as a Skolem function.

Now, let $\mathcal{L}'$ be the language of first order logic obtained from $\mathcal{L}$ by adding the predicate symbols P_G to the set of predicate symbols of $\mathcal{L}$ and the function symbols f_G to the set of function symbols of $\mathcal{L}$. For a formula G as above, assume that $x_1, \ldots, x_k$ are the variables which have free occurrences in G, and that they are given in the order of the fixed enumeration.[9] Then let L_G be the formula $P_G(x_1, \ldots, x_k)$.

Now, let us assume that F is a formula that we want to refute (the case of proving a formula valid is exactly dual to this). So we need to transform F to a formula F' in conjunctive normal form that is satisfiable iff F is satisfiable. To this end we take each of the formulas D_G defined above and transform it to clausal form using the Skolem function f_G if necessary. We thus obtain, for each subformula G of F, a formula C_G in conjunctive normal form which is defined as follows. Let us assume again that $x_1, \ldots, x_k$ are the free variables of G. Then

1. If G is an atomic formula then

$$\begin{aligned} C_G \stackrel{\text{def}}{=} & \forall x_1 \ldots \forall x_k (L_G \vee \neg G) \wedge \\ & \forall x_1 \ldots \forall x_k (\neg L_G \vee G). \end{aligned}$$

2. If $G \equiv \neg H$ then

$$\begin{aligned} C_G \stackrel{\text{def}}{=} & \forall x_1 \ldots \forall x_k (L_G \vee L_H) \wedge \\ & \forall x_1 \ldots \forall x_k (\neg L_G \vee \neg L_H). \end{aligned}$$

3. If $G \equiv H \vee I$ then

$$\begin{aligned} C_G \stackrel{\text{def}}{=} & \forall x_1 \ldots \forall x_k (L_G \vee \neg L_H) \wedge \\ & \forall x_1 \ldots \forall x_k (L_G \vee \neg L_I) \wedge \\ & \forall x_1 \ldots \forall x_k (\neg L_G \vee L_H \vee L_I). \end{aligned}$$

4. If $G \equiv H \wedge I$ then

$$\begin{aligned} C_G \stackrel{\text{def}}{=} & \forall x_1 \ldots \forall x_k (L_G \vee \neg L_H \vee \neg L_I) \wedge \\ & \forall x_1 \ldots \forall x_k (\neg L_G \vee L_H) \wedge \\ & \forall x_1 \ldots \forall x_k (\neg L_G \vee L_I). \end{aligned}$$

5. If $G \equiv H \rightarrow I$ then

$$\begin{aligned} C_G \stackrel{\text{def}}{=} & \forall x_1 \ldots \forall x_k (L_G \vee L_H) \wedge \\ & \forall x_1 \ldots \forall x_k (L_G \vee \neg L_I) \wedge \\ & \forall x_1 \ldots \forall x_k (\neg L_G \vee \neg L_H \vee L_I). \end{aligned}$$

[9] For short, we shall just say "Let $x_1, \ldots, x_k$ be the free variables of G".

6. If $G \equiv \exists x H$ then

$$C_G \stackrel{\text{def}}{=} \forall x_1 \ldots \forall x_k \forall x(L_G \vee \neg L_H) \wedge \\ \forall x_1 \ldots \forall x_k(\neg L_G \vee L)$$

where the atomic formula L is obtained from the atomic formula L_H by replacing all occurrences of x by $f_G(x_1, \ldots, x_k)$.

7. If $G \equiv \forall x H$ then

$$C_G \stackrel{\text{def}}{=} \forall x_1 \ldots \forall x_k(L_G \vee \neg L) \wedge \\ \forall x_1 \ldots \forall x_k \forall x(\neg L_G \vee L_H)$$

where the atomic formula L is obtained from the atomic formula L_H by replacing all occurrences of x by $f_G(x_1, \ldots, x_k)$.

Now, let C be the conjunction of all the formulas C_G such that G is a subformula of F.[10]

Definition 2.2.1
The (conjunctive) *definitional form* F' of F is the formula in conjunctive normal form obtained from $C \wedge L_F$ by repeated application of the associativity law for $\wedge$. We denote the set of clauses obtained from C by def_F. The clause set def_F is said to be the *definition* for F.

The structure-preserving transformation to clausal form can also be applied to formulas containing other propositional connectives than $\neg$, $\vee$, $\wedge$ and $\rightarrow$. In fact, for any fixed finite set of propositional connectives of arbitrary arities, the number of literals of the resulting formula will still be linear in the number of occurrences of propositional connectives and quantifiers in the input formula. This is important especially for the frequently occurring connective of equivalence, $\leftrightarrow$, because it is a well known fact that, even in propositional logic, for a multiple equivalence

$$A_1 \leftrightarrow (A_2 \leftrightarrow (A_3 \ldots \leftrightarrow (A_{n-1} \leftrightarrow A_n) \ldots))$$

between propositional variables $A_1, \ldots, A_n$, any formula in clausal form that is semantically equivalent to it will necessarily have at least 2^{n-1} clauses.

The idea of transforming a formula to clausal form by introducing definitions is known to logicians for a long time. Similar transformations have been used by Thoralf Skolem in the proof of the Löwenheim-Skolem theorem in [Sko20][11] and by Kurt Gödel in the proof of his completeness theorem in [Göd30]. In 1966, G.S. Tseitin introduced a structure-preserving transformation to clausal form for the case of propositional logic. An English translation of his paper can be found

[10]We can choose an arbitrary order, e.g., the lexicographic order.

[11]Actually, the structure-preserving transformation to clausal form is closer to the Skolem normal form presented in Skolem's paper than it is to the Skolem normal form that is standard today.

in [Tse70]. Stephen Cook and Robert Reckhow used a similar transformation in [CR74], [Rec76], and [CR79], as a translation between two languages of propositional logic in order to obtain simulation results for Frege systems with substitution and for extended Frege Systems. In the field of automated theorem proving, the structure-preserving transformations to normal form began to be known and applied in actual implementations in the eighties. Greenbaum, Nagasaka, Rorke and Plaisted demonstrate in [GNOP82] the structure-preserving transformation to clausal form on an example (without giving its explicit definition). A first formal definition of such a transformation for full first order logic was given in [Ede85a] where also an estimate of its complexity can be found. See also [Ede87]. Similar transformations were given by David Plaisted and Steven Greenbaum [PG86] and by D. L. Poole [Poo84]. In [BdlT89], Thierry Boy de la Tour shows that, in some cases, an even slightly better behaviour in terms of the number of clauses generated, is obtained by hybrids between the standard and the structure-preserving transformations to clausal form.

2.3 Complexity Measures for Resolution Refutations

In this book we compare different proof calculi for automated theorem proving in terms of complexities of shortest proofs. One measure of complexity is the number of proof steps where each proof step may include a unification. We shall not say very much about the complexity of one unification step itself because, where we have results of simulatability of one calculus in the other, we shall also have a rather direct correspondence between a unification step in one calculus and the corresponding unification step in a simulation in the other calculus. This may require a non-standard concept of representing unifiers such as the concept of unifier sets introduced in Section 4.1 for connection structures, but it can be done. So we shall consider a unification step as one unit in parts of this book.

Before we consider particular complexity measures we need a definition.

Definition 2.3.1
We say that (an occurrence of) a clause c in a resolution refutation $\mathcal{R}$ is *superfluous* if there does not exist a finite sequence $(c_0, \ldots, c_n)$ of occurrences of clauses in $\mathcal{R}$ such that the following three conditions hold.

1. $c_0 = c$
2. c_n is the last clause of $\mathcal{R}$, the empty clause.
3. For each $j = 1, \ldots, n$, the clause c_j is a factor of c_{j-1} or the clause c_{j-1} is one of the two parent clauses of c_j in $\mathcal{R}$.

The result of deleting all superfluous clauses from a resolution refutation of a set S of clauses is again a resolution refutation of S. We are only interested in resolution refutations without superfluous clauses here. A resolution refutation has a superfluous clause iff it has a clause which is not the last clause $\square$ of the refutation and which is not a parent clause of a resolution or factorization step in the refutation.

The first complexity measure for resolution refutations that we consider is the length of a refutation.

Definition 2.3.2
The *length* of a resolution refutation is the number of its resolution and factorization steps.

Let us see whether this is a reasonable measure of complexity for estimating the time needed for a refutation. For propositional logic this is certainly the case if we only care about complexity modulo polynomial bounds. For, the cardinality of any resolvent occurring in a refutation is always less than the total number of literals in all non-superfluous input clauses which is in turn at most twice the length of the refutation. For this reason G. S. Tseitin [Tse70] and others have used the length of a refutation as a complexity measure in propositional logic.

Now, consider first order logic. One thing that is immediately evident is that, in a resolution step, two or more literals may be merged to one due to the application of a substitution.

As an example, let n be a non-negative integer, let $x_1, \ldots, x_n$ be pairwise distinct variables, and let $a_1, \ldots, a_n$ be pairwise distinct constants. Then consider the following two clauses c_1 and c_2.

$$\begin{aligned} c_1 &\stackrel{\text{def}}{=} \{P(t_1, \ldots, t_n) \mid t_i \in \{x_i, a_i\} \quad \text{for all } i = 1, \ldots, n\} \\ c_2 &\stackrel{\text{def}}{=} \{\neg P(a_1, \ldots, a_n)\}. \end{aligned}$$

For $n = 2$, for example, we have

$$\begin{aligned} c_1 &= \{P(x_1, x_2), P(x_1, a_2), P(a_1, x_2), P(a_1, a_2)\} \\ c_2 &= \{\neg P(a_1, a_2)\} \end{aligned}$$

Resolving upon the literals $P(x_1, \ldots, x_n)$ and $\neg P(a_1, \ldots, a_n)$ yields the empty clause as a resolvent of c_1 and c_2. In this example the length of the derivation is 1, although an actual resolution theorem prover would take $2^n - 1$ steps of merging literals that have become identical through application of a substitution. A similar effect as by the resolution in this example is obtained when a factor of the clause c_1 is computed where the factorized literals of c_1 are $P(x_1, \ldots, x_n)$ and $P(a_1, \ldots, a_n)$.

You might think from this example that this problem occurred only because the number of literals in the input clauses we started from is already of the order 2^n. However, this is not the case as the following proposition shows.

Proposition 2.3.1
Let n be a non-negative integer. Then there is a set S of clauses and a resolution refutation $\mathcal{R}$ of S such that the following statements hold.

1. *The total number of occurrences of literals in S is $3n+2$, and each literal has length (number of symbols) $n+1$.*

2. *$\mathcal{R}$ has no superfluous clauses.*

3. *$\mathcal{R}$ consists of exactly $2n$ resolution steps and no factorization steps.*

4. *There is a clause occurring in $\mathcal{R}$ which has 2^n+1 literals.*

Proof: We show that the clause c_1 in the example above can itself be generated by $2n-1$ resolution steps from n clauses $d_1, \ldots, d_n$ having 3 literals each, and from one unit clause d. In order to see this, assume we have $n+1$ pairwise distinct n-ary predicate symbols $P_0, \ldots, P_n$ where P_0 is identical to the predicate symbol P above. Now, let

$$d_j \stackrel{\text{def}}{=} \{\neg P_j(x_1,\ldots,x_n), P_{j-1}(x_1,\ldots,x_n), P_{j-1}(x_1,\ldots,x_{j-1},a_j,x_{j+1},\ldots,x_n)\}$$

for $j=1,\ldots,n$, and let

$$d \stackrel{\text{def}}{=} \{P_n(x_1,\ldots,x_n)\}.$$

Further let us define

$$\begin{aligned} e_j \stackrel{\text{def}}{=} & \{\neg P_j(x_1,\ldots,x_n)\} \cup \\ & \{P(t_1,\ldots,t_j,x_{j+1},\ldots,x_n) \mid t_i \in \{x_i,a_i\} \quad \text{for all } i=1,\ldots,j\} \end{aligned}$$

for all $j=1,\ldots,n$. Then $d_1=e_1$ and, for $j=2,\ldots,n$, the clause e_j is obtained by resolving d_j with e_{j-1} and then resolving the obtained resolvent again with e_{j-1}. Finally, the clause c_1 is obtained as the resolvent of e_n and d. So let

$$S \stackrel{\text{def}}{=} \{d_1,\ldots,d_n,d,c_2\}.$$

Then the assertion of the proposition holds.

q.e.d.

So we see that the length of a resolution refutation is not sufficient as a measure of its complexity for our purposes. However, in this example, there is a different refutation with $6n+2$ occurrences of literals as can be seen easily. So the example shows only the inadequacy of this measure of complexity for some particular refutation. It does not say that it is no adequate measure for the minimal complexity of derivations for a given set of clauses. Recently it has been shown by R. Letz [Let91] that the length of the shortest refutation is not even an adequate complexity measure in this weaker sense. However, it definitely is adequate for tree resolution.

Proposition 2.3.2

Let S be a finite set of clauses, let n be the maximal cardinality of clauses of S. Let $\mathcal{R}$ be a tree resolution refutation of S whose number of resolution steps is r. Let N be the number of literal occurrences in $\mathcal{R}$. Then $N \le \frac{1}{2}n^2(r+1)^2$.

Proof: We can reorder the sequence $\mathcal{R}$ of clauses so that it assumes the form $(c_0, \ldots, c_r, d_1, \ldots, d_s)$ where each c_i is an occurrence of an element of S and each d_j is a resolvent or a factor of previous elements of $\mathcal{R}$. Note that there may be clauses of S for which there is more than one occurrence in $\mathcal{R}$. This may be necessary because $\mathcal{R}$ is to be a tree derivation. For $j = 0, \ldots, s$ let C_j be the set of clause occurrences in $\{c_0, \ldots, c_r, d_1, \ldots, d_j\}$ which have not been resolved upon or factorized in the refutation $\mathcal{R}$ up to the clause occurrence d_j. For $j = 0, \ldots, s$ let N_j be the number of literal occurrences in C_j. Then $N_j \leq N_{j-1} - 1$ and $N_0 = \sum_{i=0}^{r} |c_i| \leq n(r+1)$. Therefore $\sum_{j=0}^{s} N_j \leq \frac{1}{2} N_0(N_0 - 1) \leq \frac{1}{2} n^2 (r+1)^2$. Since each clause occurrence in $\mathcal{R}$ has not been resolved upon or factorized up to that occurrence, the assertion follows.

q.e.d.

We shall see that we can improve on this estimate slightly when we allow reordering of the proof steps in the refutation. Namely, we shall obtain a cubic bound on the number literal occurrences. First we give a few definitions.

Definition 2.3.3
Let L be an occurrence of a literal in a parent clause in a resolution refutation $\mathcal{R}$ such that L is not resolved away at the respective resolution step. Then there is an occurrence of σL in the resolvent where σ is the mgu of the resolution step, and similarly for factorization steps. This occurrence of σL is said to be an *immediate descendant* of the occurrence of L in $\mathcal{R}$. The relation '... is a *descendant* of ...' is defined to be the reflexive-transitive closure of the relation '... is an immediate descendant of ...'.

In tree resolution, each literal occurrence has at most one immediate descendant.

Definition 2.3.4
If c and d are clauses then we say that c *subsumes* d, or that d is obtained from c by *subsumption*, if there is a substitution σ such that $\sigma c \subseteq d$. A set S' of clauses is obtained from a set S of clauses by *subsumption* if $S' = S \cup \{c\}$ where c is subsumed by a clause of S. The concept of a descendant will be used also for subsumption.

Note that what we mean by a subsumption step here is different from what is usually meant by subsumption in the field of automated theorem proving. There, a subsumption step consists of removing a clause from the set of clauses S if it is subsumed by another clause of S. In contrast, we mean exactly the converse here, namely the addition of a clause c to S where c is subsumed by a clause of S.

Now, let $\mathcal{R}$ be a derivation by resolution, factorization and subsumption steps. Then two literal occurrences K and L of $\mathcal{R}$ are said to be *merged* in $\mathcal{R}$ if there is a clause c in $\mathcal{R}$ and a descendant K' of K and a descendant L' of L such that K' and L' are both in c and the following holds. $K' \equiv L'$ or c is a parent clause of a resolution step in which K' and L' are resolved away (see Andrews [And76]).

Definition 2.3.5
If, in a subsumption step yielding a clause d from a clause c, no literals are merged, we speak of a *normal subsumption* or *n-subsumption*. We say then that the clause c *n-subsumes* the clause d.

If c n-subsumes d then the cardinality of c is less than or equal to the cardinality of d. Every subsumption step can be decomposed in a number of factorization steps followed by an n-subsumption step.

Lemma 2.3.1
If a clause c subsumes a clause c', and a clause d subsumes a clause d', and e is a resolvent of c' and d', then at least one of the following three statements holds.

1. *c subsumes e.*

2. *d subsumes e.*

3. *There are clauses c'' and d'' obtained from c and d, respectively, by repeated factorization, such that e is n-subsumed by a resolvent of c'' and d''.*

Proof: If there is no immediate descendant of a literal of c in c' that is resolved upon in the given resolution step then c subsumes e. If there is no immediate descendant of a literal of d in d' that is resolved upon in the given resolution step then d subsumes e. Now let there exist immediate descendants of literals of c in c' and of d in d' that are resolved upon in the resolution step. Then let c'' be obtained from c' by factorizing all pairs of literals in c' that are merged by the given subsumption steps and by the given resolution step. Similarly, let d'' be obtained from d' by factorizing all pairs of literals in d' that are merged by the given subsumption steps and by the given resolution step. Now, let c_1 be the set of literals of c which have descendants in c' that are resolved away in building the resolvent e of c' and d'. Then there is exactly one descendant K in c'' of the literals of c_1. Let the literal L in the clause d'' be determined in the analogous way. Let e'' be the resolvent of c'' and d'' with respect to the literals K and L. Then e'' n-subsumes e.

q.e.d.

Lemma 2.3.2
Let S be a set of clauses and let $\mathcal{D}$ be a derivation of a clause c from S by resolution and subsumption. Then there is a clause d subsuming the clause c such that there is a resolution derivation $\mathcal{D}'$ of d from S. Moreover, the number of resolution steps in $\mathcal{D}'$ is at most equal to the number of resolution steps in $\mathcal{D}$. Each clause of $\mathcal{D}'$ n-subsumes some clause of $\mathcal{D}$, each resolvent in $\mathcal{D}'$ n-subsumes some resolvent in $\mathcal{D}$, and $\mathcal{D}'$ contains no superfluous clauses.

Proof: This follows immediately from the previous lemma by induction on the number of resolution steps in $\mathcal{D}$.

q.e.d.

Proposition 2.3.3
Let S be a finite set of clauses, let k be the maximum of the cardinalities of the clauses of S, and let $\mathcal{R}$ be a tree resolution refutation of S with exactly n resolution steps (and an arbitrary number of factorization steps). Then there is a tree resolution refutation of S with less than $4n^3 + nk^2$ literal occurrences.

Proof: If c, d and e are clauses then let us say that e is obtained from c and d by rule R if there is a resolvent e' of c and d such that e is obtained from e' by subsumption. We say that e is obtained from c and d by rule R' if, in addition, every literal of e is a descendant of at most one literal of c and of at most one literal of d, and if, moreover, from each of the clauses c and d, only one literal is resolved away. If e is obtained from c and d by rule R' then $|c| \leq |e| + 1$ and $|d| \leq |e| + 1$.

We shall now move the factorization steps in the refutation $\mathcal{R}$ as far towards the beginning of $\mathcal{R}$ as possible. This will result in an R'-refutation $\mathcal{R}'$ of S' with at most n steps, where S' is a set of clauses obtained from clauses of S by repeated applications of factorization steps. The cardinalities of the clauses of $\mathcal{R}'$ are at most n by the above inequalities. Finally, we shall transform $\mathcal{R}'$ to a resolution refutation of S with the required properties.

First, note that if a clause d is obtained from a clause c by repeated application of factorization then it is obtained from c by subsumption. So there is a set S_1 of clauses obtained from clauses of S by repeated factorizations and an R-refutation $\mathcal{R}_1$ of S_1 in n steps. Second, note that if a clause e is obtained from clauses c and d by rule R then there are clauses c' and d' obtained from c and d, respectively, by repeated factorization, such that e is obtained from c' and d' by rule R'. We just have to factorize in c all pairs of literals which are merged in the application of R, and similarly for d. Since an application of rule R followed by a factorization is again an application of rule R, we can transform[12] the R-refutation $\mathcal{R}_1$ to an R'-refutation $\mathcal{R}'$ of a set S' of clauses obtained from clauses of S by repeated factorizations. Moreover, $\mathcal{R}'$ has exactly n steps. The easy proof is by induction on the length of $\mathcal{R}_1$.

Now, let us consider a derivation step in $\mathcal{R}'$ deriving a clause e from two clauses c and d by rule R'. Then, by definition of the rule R', there is a resolvent e' of c and d such that e' subsumes e. For each such derivation step, let us insert the resolvent e' in the derivation $\mathcal{R}'$ so that we obtain a refutation $\mathcal{R}''$ of S' by resolution and subsumption steps. Since each clause of $\mathcal{R}'$ has at most n literals, the resolvents have at most $2n-2$ literals. Now, let $\bar{\mathcal{R}}$ be the resolution refutation of S' obtained from $\mathcal{R}''$ by the construction in the proof of Lemma 2.3.2. By

[12]Just start at the end of $\mathcal{R}_1$.

Lemma 2.3.2, the number of resolution steps in $\bar{\mathcal{R}}$ is at most n, and since each resolvent in $\bar{\mathcal{R}}$ n-subsumes a resolvent in $\mathcal{R}''$, it follows that the resolvents in $\bar{\mathcal{R}}$ have cardinalities $\leq 2n-2$. So all clauses of $\bar{\mathcal{R}}$ — except those obtained by repeated factorization from clauses of S — have cardinalities $\leq 2n-2$. Moreover, at most $2n-3$ factorization steps can be applied to any clause of $\bar{\mathcal{R}}$ before a resolution step is applied. So the number of these clauses is at most $n(2n-2)$ because there are at most n resolution steps. The total number of occurrences of literals in them is therefore at most $(2n-2)^2 n$. The number of occurrences of literals in a derivation of a clause c from some clause of S by repeated application of factorization is at most $\frac{1}{2}k(k-1)$. This number has to be multiplied by the number of leaves of the refutation tree which is at most $n+1$. So the total number of occurrences of literals in the constructed tree resolution refutation of S is at most $\frac{1}{2}k(k-1)(n+1)+(2n-2)^2 n$ which is certainly less than $4n^3+nk^2$.

q.e.d.

Proposition 2.3.4
Let S be a finite set of clauses, let k be the maximum of the cardinalities of the clauses of S, and let $\mathcal{R}$ be a resolution refutation of S whose number of resolution steps is n. Then there is a resolution refutation $\mathcal{R}'$ of S such that the clauses of $\mathcal{R}'$ have cardinalities less than or equal to $\max(k,n)$ *and such that the depth of the tree corresponding to $\mathcal{R}'$ is less than or equal to $k+n$.*

Proof: First, the given refutation $\mathcal{R}$ is expanded to a tree refutation. The depth of the corresponding tree with respect to resolution steps is then at most n. Then the same construction as in the proof of the last proposition is made to obtain $\mathcal{R}'$. The proof of the assertion is then straight forward and will not be given here.

q.e.d.

This proposition shows that we can keep the cardinalities of the clauses and the depth of the proof tree small, but we may have to pay for this by having to expand the given refutation to a tree refutation which may take exponential time and space.

I do not know whether a similar construction can be made without having to expand the refutation to a tree refutation. Anyhow, the example above shows that we have to take into account the length of a single resolution step as well as the number of resolution steps, at least if we are interested in a complexity measure for any particular refutations (not only for shortest ones).

It is known that even a single resolution step can give rise to an exponential explosion of the number of symbol occurrences in the resolvent. This is due to the unification process which is exponential when terms are written in their standard representation. There is, however, a way out of this problem. The terms have to be represented so that multiply occurring subterms are represented only once, by some sort of dag representation of terms. This is done in the linear or almost linear unification algorithms by Huet [Hue76], Martelli and Montanari [MM82],

Paterson and Wegman [PW78] and others. In Section 4.1 we shall introduce a representation of unifiers suited to our purposes that also has this property.

Note, however, that in a sequence of resolution steps there can be an exponential increase of the length of the unifier even in such a dag representation. Take, for example, the clause $c_0 \stackrel{\text{def}}{=} \{\neg Px, Pfx\}$ and let c_{n+1} be the resolvent of the clause c_n with itself for each nonnegative integer n. Such an exponential explosion of the length of the unifier cannot happen in the connection calculus (without splitting) since derivations containing such an exponential increase of lengths of unifiers cannot be expressed in the connection calculus. There the new mgu is obtained as a merge of the old one with the mgu of some link. The maximal length of an mgu of a link is at most of the order of the size of the input formula, and the length of the merge of two substitutions is at most the sum of the lengths of the two substitutions when a dag representation is chosen for them. So there can only be a linear increase of lengths of substitutions in the connection calculus. We shall see in Section 2.5 that there are also resolution refutations without such an exponential increase that cannot be expressed in the connection calculus.

We could have defined the clauses to be multisets of literals instead of sets of literals. Then we could have defined a factorization step to remove one of the two factorized literals from the clause (and to apply the unifier to the clause as before). This would be a definition that is closer to actual implementation. If $|c|$ denotes the cardinality of a multiset c then the following would hold. If d is a factor of c then $|d| = |c| - 1$, and if d is a resolvent of c_1 and c_2 then $|d| \geq |c_1| - 1$ and $|d| \geq |c_2| - 1$. This implies that in a resolution refutation of length n every clause which is not superfluous has at most n literals, and thus the total number of literals of the refutation is at most n^2. Thus the length of a refutation is a more adequate measure of complexity in this setting.

Another measure of complexity for resolution refutations is the number of resolution steps together with the maximum number of literals of clauses in the refutation.

Proposition 2.3.5

Let S be a set of clauses and let $\mathcal{R}$ be a resolution refutation of S without superfluous clauses. Let n be the number of resolution steps in $\mathcal{R}$, and let m be the maximal cardinality of clauses occurring in $\mathcal{R}$. Then the total number of factorization steps in $\mathcal{R}$ is at most $2mn$. The total number of occurrences of literals in $\mathcal{R}$ is at most $2m^2n$. The same holds for resolution on multisets of literals.

Proof: Since each clause can have at most m literals, factorization can only be applied $m-1$ times to any clause before it is used as a parent clause in a resolution step. So, for every resolution step, the total number of factorization steps applied to yield its resolvent is at most $2(m-1)$, and thus the total number of factorization steps in $\mathcal{R}$ is at most $2(m-1)n$. Thus the total number of inference steps is at most $(2m-1)n$, and the total number of non-empty clauses of $\mathcal{R}$ is at most $(2m-1)n$.

q.e.d.

2.4 Simulation of the Connection Calculus by Resolution

In this section we show that any proof in the connection calculus described in Section 1.3 can be simulated in resolution step by step in the sense that to every chosen connection in an extension step corresponds exactly one resolution step.

In order to do this we associate a set of literals with each structured matrix. This set of literals will then be a clause of the corresponding resolution refutation, or will at least be subsumed by such a clause. The corresponding resolution refutation can then be constructed immediately.

Definition 2.4.1
Let $S = (M, p, D, \sigma)$ be a structured matrix in the connection method. Then the *set of literals associated* with S is the set $c[S] \stackrel{\text{def}}{=} \sigma(\bigcup D)$, i.e., the set of all literals obtained by applying the current substitution to the unsolved subgoals. The *sequence of clauses associated* with a derivation of a matrix in the connection method is the sequence of the sets of literals associated with the structured matrices of the derivation.

Proposition 2.4.1
Let M be a matrix and let Δ be a derivation of M in the connection method. Let m be the number of choices of connections done in Δ. Then there is a resolution refutation $\mathcal{R}$ of the matrix M which has at most m resolution steps. Moreover, the maximum cardinality of clauses of $\mathcal{R}$ is less than or equal to the maximum number of unsolved subgoals in structured matrices of Δ plus the maximum of the cardinalities of the clauses of M.

In order to prove Proposition 2.4.1, we first consider a single extension step in the connection proof Δ. The proof of the following lemma shows how such an extension step can be simulated by resolution and subsumption. We shall then use the lemma to simulate the whole connection proof by resolution and subsumption. The resulting derivation can finally be transformed to a resolution derivation.

Lemma 2.4.1
Let M be a matrix and let $\Delta = (S_0, \ldots, S_r)$ be a connection derivation of M and $s \leq r$. Assume that the derivation step from S_{s-1} to S_s is an extension step, and let n be the number of choices of connections done in this step. Then there is a sequence $(e_1, \ldots, e_n)$ of n clauses such that $e_n = c[S_s]$ and, for all $m = 1, \ldots, n$, the clause e_m is obtained by resolution from two clauses each of which is subsumed by a clause of $M \cup \{c[S_0], \ldots, c[S_{s-1}]\} \cup \{e_1, \ldots, e_{m-1}\}$. Moreover, each e_m is an instance of a subset of the union of the set of unsolved subgoals of S_s with the clause chosen in the extension step leading from S_{s-1} to S_s.

Proof: Let $S_s = (M, (L_1, \ldots, L_k), (d_1, \ldots, d_{k+1}), \sigma)$. For each of the literals L_j of the active path there is some S_t with $t < s$ which has the form

$$S_t = (M, (L_1, \ldots, L_{j-1}), (d_1, \ldots, d_{j-1}, d_j \cup \{L_j\}), \tau).$$

Let c_j be the set of literals associated with it. Now let c be the clause extended to in the step from S_{s-1} to S_s. Now let $L_{j_1},\ldots,L_{j_n}$ be the literals of the active path of S_s to which connections have been chosen in the derivation step from S_{s-1} to S_s, and let $K_{j_1},\ldots,K_{j_n}$ be the corresponding complementary literals in the clause c. We define a sequence $(e_0,\ldots,e_n)$ of sets of literals: Let

$$e_m \stackrel{\text{def}}{=} \sigma((c \setminus \{K_{j_1},\cdots,K_{j_m}\}) \cup d_1 \cup \ldots \cup d_k) \qquad \text{for} \quad m \leq n.$$

Then each e_m can be obtained from e_{m-1} and σc_{j_m} by one resolution step. Moreover, c subsumes e_0, each σc_{j_m} is subsumed by some $c[S_t]$ with $t < s$, and $e_n = c[S_s]$.

q.e.d.

We illustrate this proof with an example. Let us assume that M is a matrix containing all clauses shown in Figure 2.1 (and possibly some more clauses). Assume that S_{s-1} is the structured matrix $(M,(L_1,L_2,L_3),(d_1,d_2,d_3,d_4'),\rho)$ shown in Figure 2.1.

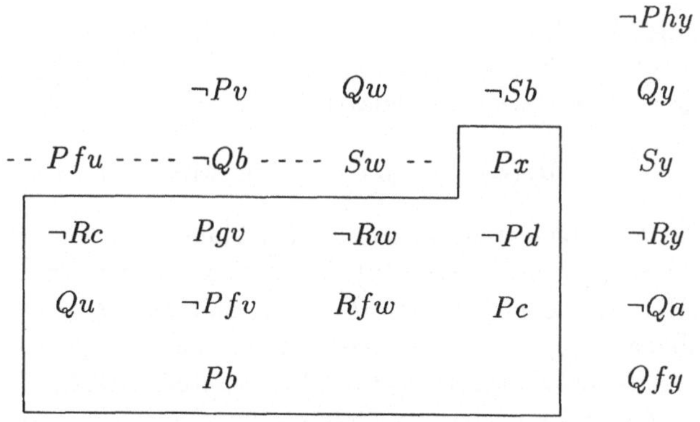

				$\neg Phy$
	$\neg Pv$	Qw	$\neg Sb$	Qy
Pfu	$\neg Qb$	Sw	Px	Sy
$\neg Rc$	Pgv	$\neg Rw$	$\neg Pd$	$\neg Ry$
Qu	$\neg Pfv$	Rfw	Pc	$\neg Qa$
	Pb			Qfy

$$\rho = \{v \leftarrow fu, w \leftarrow b\}.$$

Figure 2.1: The structured matrix S_{s-1}

In Figure 2.1, the active path (L_1,L_2,L_3) has been indicated by a dashed line. So $L_1 \equiv Pfu$, $L_2 \equiv \neg Qb$, and $L_3 \equiv Sw$. The unsolved subgoals are the literals enclosed in the box. So, $d_1 = \{\neg Rc, Qu\}$, $d_2 = \{Pgv, \neg Pfv, Pb\}$, $d_3 = \{\neg Rw, Rfw\}$, and $d_4' = \{Px, \neg Pd, Pc\}$. We use here the notation in the proof of Lemma 2.4.1. The set $c_4 = c[S_{s-1}]$ of literals associated with the structured matrix S_{s-1} is the set of literals enclosed in the box (its unsolved subgoals), with the current substitution ρ applied to them. This structured matrix may, for example,

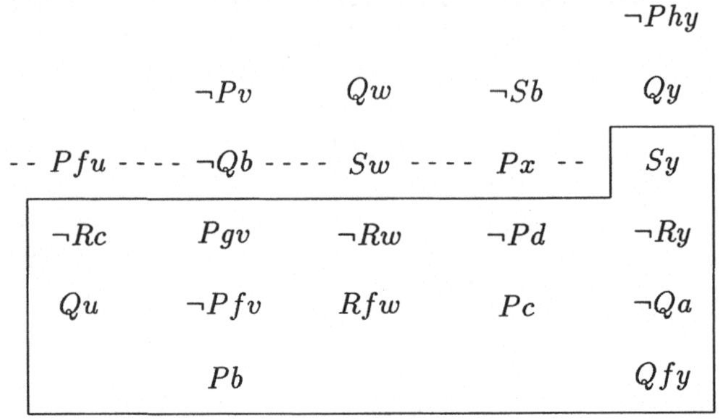

				$\neg Phy$
	$\neg Pv$	Qw	$\neg Sb$	Qy
-- Pfu ----	$\neg Qb$ ----	Sw ----	Px --	Sy
$\neg Rc$	Pgv	$\neg Rw$	$\neg Pd$	$\neg Ry$
Qu	$\neg Pfv$	Rfw	Pc	$\neg Qa$
	Pb			Qfy

$$\sigma = \{v \leftarrow fu, w \leftarrow b, x \leftarrow hb, y \leftarrow b\}$$

Figure 2.2: The structured matrix S_s

have been obtained by three extension steps using the connections $\{Pfu, \neg Pv\}$, $\{\neg Qb, Qw\}$, and $\{Sw, \neg Sb\}$, respectively.

Now, assume that the structured matrix

$$S_s = (M, (L_1, L_2, L_3, L_4), (d_1, d_2, d_3, d_4, d_5), \sigma)$$

is obtained from S_{s-1} by extension to the last clause c in Figure 2.1, with the connections $\{Px, \neg Phy\}$ and $\{\neg Qb, Qy\}$. Then S_s is the connected matrix shown in Figure 2.2. Here, $L_4 \equiv Px$, $d_4 = \{\neg Pd, Pc\}$, and $d_5 = \{Sy, \neg Ry, \neg Qa, Qfy\}$. Its unsolved subgoals are, again, enclosed in a box, and applying the substitution σ to them yields the set of literals $c_5 = c[S_s]$ associated with S_s.

Now, Figure 2.3 shows an earlier stage of the connection proof at which the extension to the second clause has just been made. The structured matrix depicted in Figure 2.3 is some structured matrix S_t with $t < s$. The set of literals associated to it is $c_2 = c[S_t]$, in the notation of the proof of Lemma 2.4.1.

Now, the set $c_5 = c[S_s]$ of literals associated with S_s (see Figure 2.2) can be obtained in two resolution steps from the clause c of M and from the sets c_2 and c_4 of literals associated with the structured matrices S_t and S_{s-1}, respectively. Each of these resolution steps corresponds to one of the two connections $\{Px, \neg Phy\}$ and $\{\neg Qb, Qy\}$ chosen in the extension step. In Figure 2.4, the sets of unsolved subgoals of S_t and S_{s-1}, and the clause c, are indicated by boxes drawn with thin lines. The set of unsolved subgoals of S_s is indicated by a box drawn with thick lines. The set c_2 of literals associated with S_t is obtained from the set of unsolved subgoals of S_t by applying the substitution τ, the set c_4 of literals associated with S_{s-1} is obtained from the set of unsolved subgoals of S_{s-1} by applying the

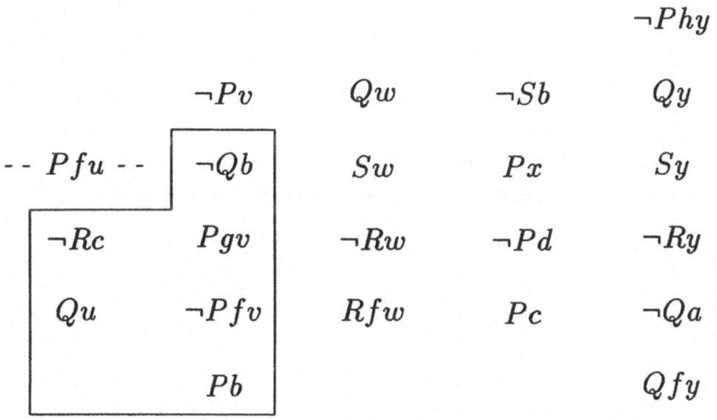

				$\neg Phy$
	$\neg Pv$	Qw	$\neg Sb$	Qy
-- Pfu --	$\neg Qb$	Sw	Px	Sy
$\neg Rc$	Pgv	$\neg Rw$	$\neg Pd$	$\neg Ry$
Qu	$\neg Pfv$	Rfw	Pc	$\neg Qa$
	Pb			Qfy

$$\tau = \{v \leftarrow fu\}$$

Figure 2.3: The structured matrix S_t

substitution ρ, and the set c_5 of literals associated with S_s is obtained from the set of unsolved subgoals of S_s by applying the substitution σ. By resolving the three sets of literals thus corresponding to the three thin boxes in Figure 2.4, we obtain the set of literals corresponding to the thick box. In Figure 2.4, the connections chosen in the extension step, are shown. It is now easily seen that a resolution of c with c_4 (using the connection $\{Px, \neg Phy\}$ in Figure 2.4) and a subsequent resolution of the obtained resolvent with c_2 (using the connection $\{\neg Qb, Qy\}$ in Figure 2.4) yields the clause c_5 of Figure 2.2.

Proof of Proposition 2.4.1: Now, let the inference calculus C be defined as resolution, except that, instead of the factorization rule we have the rule

$$c \vdash d \quad \text{if } c \text{ subsumes } d.$$

Then, it follows from Lemma 2.4.1 that, under the assumptions of Proposition 2.4.1, there is a refutation of M in the calculus C using at most m resolution steps (and an arbitrary number of subsumptions), and for the cardinalities of the clauses of this refutation, the estimate Lemma 2.4.1 holds. In fact, in the case of a truncation step, the associated set of literals does not change. And in the case of a factorization step of the connection method, the new associated set of literals is obtained from the old one just by factorization in the resolution calculus. Now, subsumption steps can be replaced by factorization steps, which proves the proposition.

q.e.d.

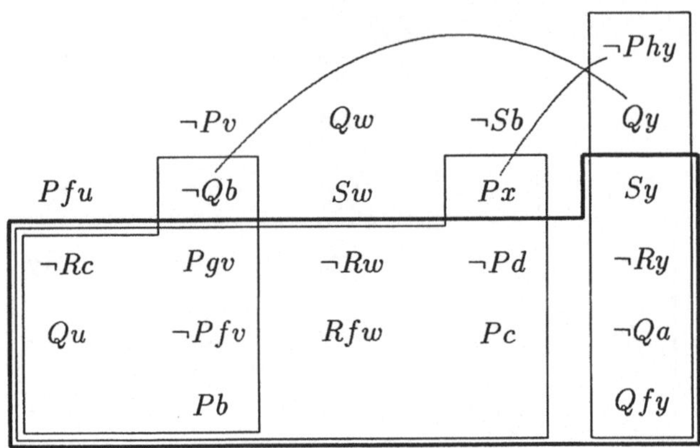

$$\tau = \{v \leftarrow fu\} \qquad \rho = \{v \leftarrow fu, w \leftarrow b\}$$

Figure 2.4: Simulation of an extension step by resolution

Theorem 2.4.1
Resolution can p-simulate the connection calculus.

Proof: Assume we are given a connection proof Δ for a set of clauses M. Let $\mathcal{R}'$ be a resolution refutation of M according to Proposition 2.4.1. Let $\mathcal{R}$ be the resolution refutation obtained from $\mathcal{R}$ by deleting all superfluous clauses. From Propositions 2.4.1 and 2.3.5, it follows that there is an upper bound for the number of occurrences of literals in $\mathcal{R}$ which is polynomial in the size of Δ. Since the applied substitution σ in the proof of Lemma 2.4.1 is the current substitution of the structured matrix S_s of the derivation Δ, and since a similar statement holds for the simulation of factorization steps of the connection proof, the maximal size of the literalas of $\mathcal{R}$ also has an upper bound which is polynomial in the size of Δ. So the size of $\mathcal{R}$ has an upper bound which is polynomial in the size of Δ.

q.e.d.

There are also other ways of mapping connection proofs to resolution refutations. But not all of them preserve the time order, which means that we do not have a simulation in the strict sense for some of them.

2.5 Non-Simulatability of Resolution in the Connection Calculus

In this section we give examples of formulas which show that we cannot in general p-simulate resolution refutations in the version of the connection calculus given here. In fact, the number of derivation steps increases exponentially in the worst case when going from the resolution method to the connection method.

As a first example we take the formula

$$Pa \wedge \forall x(Px \rightarrow Pfx) \rightarrow Pffffffffa$$

where P is a predicate symbol and f a function symbol both of arity 1, and where we have omitted parentheses wherever this does not lead to ambiguities. We shall adopt the notation $f^n t$ for $\underbrace{f \ldots f}_{n \text{ times}} t$ for any nonnegative integer n and any term t. Then the formula has the matrix representation

$$\begin{matrix} & Px & \\ \neg Pa & & Pf^8 a \\ & \neg Pfx & \end{matrix}$$

There exists a short resolution proof of this formula: Resolving the second clause with itself we obtain the resolvent $\{Px, \neg Pf^2 x\}$. Resolving this clause with itself we obtain $\{Px, \neg Pf^4 x\}$, and resolving this clause with itself we obtain $\{Px, \neg Pf^8 x\}$. By resolution with the first and the third clause we get the empty clause. As we shall see, however, the shortest proof in the connection calculus takes 9 extension steps. In general, for the matrix

$$\begin{matrix} & Px & \\ \neg Pa & & Pf^n a \\ & \neg Pfx & \end{matrix}$$

there is a resolution refutation in $2k$ steps where k is the smallest integer such that $2^k \geq n$, whereas the shortest connection derivation has $n+1$ extension steps.

In order to prove that there is no shorter connection derivation we define the *degree* of a valid matrix M to be the cardinality of the smallest (in terms of cardinality) set of variants of clauses of M which admits a unifiable spanning set of connections. Equivalently, the degree of M is the cardinality of the smallest set N of ground instances of clauses of M such that N is valid. Of course, N can be considered as a formula of propositional logic since it has no variables. The proof of the following proposition is obvious.

Proposition 2.5.1
Let M be a matrix of degree d and let Δ be a connection derivation of M. Then the number of extension steps that Δ contains is greater than or equal to $d-1$.

It is easily seen that the degree of the matrix

$$\{\{\neg Pa\}, \{Px, \neg Pfx\}, \{Pf^n a\}\}$$

is $n+2$. Thus any connection derivation of it must have at least $n+1$ extension steps.

It might seem that the increase of the length of the proof is not so severe when one considers the lengths of the occurring literals as well as the number of proof steps. In fact, the lengths of the literals are in the order of n in our example. So we define

Definition 2.5.1
The *length* of a literal is the number of nodes of its structure tree. The *complexity* of a clause is the sum of the lengths of its literals. The *complexity* of a matrix or of a resolution refutation is the sum of the complexities of its clauses.

There are, however, examples which are similar in structure to the one given above which resist this objection. Take, for example, the matrix

$$\begin{array}{ccc} & Px_1 \dots x_m & \\ \neg Pa_1 \dots a_m & & Pa_{\pi^{-1}(1)} \dots a_{\pi^{-1}(m)} \\ & \neg Px_{\pi(1)} \dots x_{\pi(m)} & \end{array}$$

where P is a predicate symbol of arity m, and π is a permutation of the set $\{1, \dots, m\}$. If we set $n \stackrel{\text{def}}{=} \text{ord}(\pi) - 1$ then we have exactly the same situation with respect to shortest derivations as in the previous example. However, in this example the lengths of the literals are in the order of m instead of n. By a suitable choice of π we obtain

Proposition 2.5.2
There is an infinite sequence $(M_1, M_2, \dots)$ of complementary matrices such that the complexity of M_k as well as the complexity of its shortest resolution refutation grow polynomially with respect to k but the number of extension steps of the shortest connection derivation grows at least exponentially with respect to k.

Proof: For each nonnegative integer j, let p_j be the sum of the first j prime numbers. We consider permutations in the symmetric group S_{p_k}, i.e., permutations of the first p_k positive integers. For positive integers $j \le k$, let π_j be the cyclic permutation

$$(p_{j-1}{+}1 \quad p_{j-1}{+}2 \quad \dots \quad p_j{-}1 \quad p_j).$$

in S_{p_k}. Then $\text{ord}(\pi_j)$ is the j-th prime number. Let

$$\pi \stackrel{\text{def}}{=} \prod_{j=1}^{k} \pi_j.$$

Then $\text{ord}(\pi)$ is the product of the first k prime numbers which grows more than exponentially with respect to k. The complexity of the matrix, however, is of the order of the sum of the first k prime numbers which grows only polynomially with respect to k.

q.e.d.

As a corollary, we have the following theorem.

Theorem 2.5.1
The connection calculus cannot p-simulate resolution.

One reason why resolution proofs cannot be p-simulated by connection derivations in the calculus given here is that resolution can make repeated use of lemmata generated in the course of the deduction. In our first example these are the clauses $\{Px, \neg Pf^j x\}$ for $j = 2,4,8,\ldots$ Another, more practical, example for the use of lemmata in resolution, namely the simplest Prolog program for computing the n-th Fibonacci number for some given nonnegative integer n, can be found in [Ede87]. It is important to note here that the use of lemmata is closely connected to the dag structure of resolution derivations. If we would expand the derivation dag into a tree then this source of efficiency of resolution would be lost. In the calculus of the connection method presented here there exist only very limited possibilities for use of lemmata. The concept of splitting introduced by W. Bibel [Bib87] expands these possibilities greatly when used in a nested way. It is, however, not at all obvious how to define a concept of nested splitting that would be sound and that would incorporate the possibilities of use of lemmata offered by resolution. The concept of connection structures presented in Chapter 4 of this book, can be regarded as such a concept, and as a generalization and extension of the concept of splitting. A first step towards the use of lemmata in actual implementations of the connection method has been done in the theorem prover SETHEO [BEK+87]. This implementation is restricted, however, to lemmata consisting of just one literal.

Another source of inefficiency of the connection calculus becomes apparent with the following example. Suppose we are given a matrix

$$\begin{matrix} & \neg Py & \\ Px & & \ldots \\ & Q & \end{matrix}$$

where the predicate symbol P does not occur in the rest of the matrix. In resolution we can resolve the first two clauses to obtain the clause $\{Q\}$ which does not contain x and y any more. In fact, a resolution prover will apply subsumption and pure clause elimination and just throw the first two clauses away. So resolution has the ability to just forget the variables x and y. On the other hand, a theorem prover using the connection calculus introduced here will generate a current substitution $\{y \leftarrow x\}$ when encountering the connection $\{Px, \neg Py\}$, and henceforth x and y will be shared variables, and the binding will be carried along all the time in the current substitution. Our previous two examples also show the necessity for forgetting of variables, here in connection with the use of lemmata. In these examples, the use of lemmata is only possible efficiently if the variables that do not occur in the formulation of a lemma, are forgotten. Otherwise, we have again an exponential explosion of the size of the derivation.

2.6 Variants of the Tableau Calculus TC

In this section we look at the application of the method of tableaux to automated theorem proving and introduce some variants of the tableau calculus TC which are more adequate for this purpose. We introduce variants of the tableaux calculus TC described in Section 1.5. The objectives of this endeavour are

1. to obtain a more natural treatment of multiple or n-ary logical connectives $\vee$ and $\wedge$ than in TC,

2. to introduce the concept of unification to the method of tableaux,

3. to study the method of tableaux in the case of formulas in clausal form, and

4. to show the close relationship between the method of tableaux and the connection method.

2.6.1 The Tableau Calculi TC_{ac}, TC_{iac} and TC_m

Let us first look at the language of first order predicate logic and at the language of some calculi used for automated theorem proving, such as the resolution method and the connection method.

Definition 2.6.1
By $\mathcal{L}$ we denote the language of first order predicate logic, i.e., the set of its formulas, considered as a free term algebra over the logical symbols $\neg$, $\vee$, $\wedge$, $\exists$, $\forall$ and over the variables, the constants, the function symbols, and the predicate symbols.

The constructors $\vee$ and $\wedge$ are binary functions mapping two formulas A and B of $\mathcal{L}$ to a formula $A \vee B$ and $A \wedge B$, respectively. They are not idempotent, commutative or associative. The formulas $A \vee B$ and $B \vee A$ are non-identical formulas in general, although they are, of course, semantically equivalent to each other. The same holds for $(A \vee B) \vee C$ and $A \vee (B \vee C)$. For the tableau calculus TC introduced in Section 1.5 this has the effect that the β-rule cannot be applied to a formula $(A \vee B) \vee C$ to split it in the two formulas A and $B \vee C$.

In most typical theorem proving methods such as the resolution or the connection method, however, multiple disjunctions $A_1 \vee \cdots \vee A_n$ or conjunctions $A_1 \wedge \cdots \wedge A_n$ are allowed. Such a disjunction or conjunction is represented as the set of its disjuncts or conjuncts, respectively. So we have a non-free term algebra in this case where the constructors $\vee$ and $\wedge$ are idempotent, associative and commutative. We shall now define a tableau calculus TC_{ac} for associative and commutative $\vee$ and $\wedge$ and a tableau calculus TC_{iac} for idempotent, associative and commutative $\vee$ and $\wedge$.

Definition 2.6.2
By $\mathcal{L}_{ac}$ we denote the non-free term algebra obtained from $\mathcal{L}$ by building its quotient with respect to associativity and commutativity of $\vee$ and $\wedge$, i.e., with respect to the smallest equivalence relation $\sim$ on $\mathcal{L}$ such that the constructors of $\mathcal{L}$ are compatible with $\sim$, and $\vee$ and $\wedge$ are associative and commutative modulo $\sim$. By $\mathcal{L}_{iac}$ we denote the non-free term algebra obtained from $\mathcal{L}$ by building its quotient with respect to idempotence, associativity and commutativity of $\vee$ and $\wedge$.

Definition 2.6.3
The tableau calculi TC_{ac} and TC_{iac} are defined exactly as the calculus TC, except that the language of formulas for TC_{ac} and for TC_{iac} is not $\mathcal{L}$ but $\mathcal{L}_{ac}$ and $\mathcal{L}_{iac}$, respectively.

In practice, the only difference between the three calculi TC, TC_{ac} and TC_{iac} is that in TC_{ac} and in TC_{iac} the α-rule and the β-rule will allow splitting a multiple conjunction or disjunction, respectively, in two parts in an arbitrary way (due to the commutativity and the associativity), and, in the case of TC_{iac}, also to omit one member of a pair of identical conjuncts or disjuncts (due to the idempotence).

Every refutation of a set S of formulas in TC is essentially also a refutation of S in TC_{iac}, but not vice versa. More specifically, for every refutation $\mathcal{T}$ of S in TC the tree $\mathcal{T}'$ is a refutation of S' in TC_{iac} where $\mathcal{T}'$ and S' are the tableau and the set of formulas in the language of TC_{iac} obtained from $\mathcal{T}$ and S, respectively, by replacing each formula in the language $\mathcal{L}$ by the corresponding formula in $\mathcal{L}_{iac}$. It follows that TC_{iac} is complete. The proof of the soundness of TC can be directly carried over to yield a proof of the soundness of TC_{iac}. The same holds for the calculus TC_{ac}.

Another way to obtain a more natural treatment of multiple conjunctions or disjunctions than in the tableau calculus TC is to introduce n-ary propositional connectives $\wedge$ and $\vee$ for every non-negative integer n. Correspondingly, α-formulas α have the form $F_1 \wedge \cdots \wedge F_n$ or $\neg(F_1 \vee \cdots \vee F_n)$. For such a formula α we define formulas $\alpha_1, \ldots, \alpha_n$ by $\alpha_i \stackrel{\text{def}}{=} F_i$ or $\alpha_i \stackrel{\text{def}}{=} \neg F_i$, respectively, for $i = 1, \ldots, n$. The corresponding definitions for β-formulas are analogous.

Definition 2.6.4
The tableau calculus TC_m for first order predicate logic with multiple arity conjunctions and disjunctions is defined exactly as the tableau calculus TC except that the α- and β-rules of TC are replaced by

α-RULE $$\frac{\alpha}{\alpha_i} \quad \text{for } i = 1, \ldots, n$$

β-RULE $$\frac{\beta}{\beta_1 \mid \cdots \mid \beta_n}$$

where $n = n[\alpha]$ or $n = n[\beta]$ is the number conjuncts or disjuncts of α or β, respectively, or the number of disjuncts or conjuncts of F if $\alpha \equiv \neg F$ or $\beta \equiv \neg F$, respectively.

We shall see that there is a close similarity between TC_m and the connection method. Therefore we shall develop the calculus TC_m to make it more suited to automated theorem proving, and we shall not say very much about the other tableau calculi introduced here. Let us make just a few remarks on the question of direct simulatability of one of the introduced tableau calculi in another one.

2.6.2 Comparison of Tableau Calculi

We have seen already that the language of TC can be translated directly to the language of TC_{ac} and to the language of TC_{iac} and that every refutation in TC can be translated to a refutation in TC_{ac} and to a refutation in TC_{iac} without increasing the number of nodes or branches of the refutation tree.

We can also translate the language of TC_m to the language of TC, TC_{ac} or TC_{iac}. We just define the translation A' of a formula A inductively by

1. If A is an atomic formula then $A' \stackrel{\text{def}}{=} A$.
2. If $A \equiv \neg B$ then $A \stackrel{\text{def}}{=} \neg B'$.
3. If $A \equiv A_1 \wedge \cdots \wedge A_n$ then $A' \stackrel{\text{def}}{=} A_1' \wedge (A_2' \wedge (\ldots (A_{n-1}' \wedge A_n') \ldots))$.
4. If $A \equiv A_1 \vee \cdots \vee A_n$ then $A' \stackrel{\text{def}}{=} A_1' \vee (A_2' \vee (\ldots (A_{n-1}' \vee A_n') \ldots))$.
5. If $A \equiv \forall x B$ then $A' \stackrel{\text{def}}{=} \forall x B'$.
6. If $A \equiv \exists x B$ then $A' \stackrel{\text{def}}{=} \exists x B'$.

Proposition 2.6.1
Let $\mathcal{T}$ be a refutation of a set S of formulas in TC_m and let C be any one of the calculi TC, TC_{ac} and TC_{iac}. Let S' be the set of translations of the formulas of S to the language of C. Then there exists a refutation $\mathcal{T}'$ of S' in the calculus C such that the number of branches of $\mathcal{T}'$ equals the number of branches of $\mathcal{T}$.

Proof: For an α-formula α of TC_m and for $n \stackrel{\text{def}}{=} n[\alpha]$ we define TC_m-formulas α_i' for $i = 1, \ldots, n$ as follows.

1. If $\alpha \equiv F_1 \wedge \cdots \wedge F_n$ then $\alpha_i' \stackrel{\text{def}}{=} F_i \wedge \cdots \wedge F_n$.
2. If $\alpha \equiv \neg(F_1 \vee \cdots \vee F_n)$ then $\alpha_i' \stackrel{\text{def}}{=} \neg(F_i \vee \cdots \vee F_n)$.

For a β-formula β of TC_m and for $n \stackrel{\text{def}}{=} n[\beta]$ we define TC_m-formulas β_i' for $i = 1, \ldots, n$ as follows.

1. If $\beta \equiv F_1 \vee \cdots \vee F_n$ then $\beta_i' \stackrel{\text{def}}{=} F_i \vee \cdots \vee F_n$.
2. If $\beta \equiv \neg(F_1 \wedge \cdots \wedge F_n)$ then $\beta_i' \stackrel{\text{def}}{=} \neg(F_i \wedge \cdots \wedge F_n)$.

In particular, we always have $\alpha_1' \equiv \alpha$ and $\beta_1' \equiv \beta$.

Let us now consider, for all α-formulas α of TC_m the rule

$$\frac{\alpha}{\alpha_1} \quad \text{and} \quad \frac{\alpha}{\alpha_2'}$$

and, for all β-formulas β of TC_m the rule

$$\frac{\beta}{\beta_1 \mid \beta_2'}.$$

We make a stepwise transformation of $\mathcal{T}$ to a tree built up according to these rules instead of the α- and β-rule of TC_m. First we replace any application of the α-rule,

$$\frac{\alpha}{\alpha_i},$$

with itself in the cases of TC_{ac} and of TC_{iac}, and with a sequence

$$\frac{\alpha_1'}{\alpha_2'}$$

$$\frac{\alpha_2'}{\alpha_3'}$$

$$\vdots$$

$$\frac{\alpha_i'}{\alpha_i}$$

of applications of the first of the two rules given above, in the case of TC. We replace any application of the β-rule,

$$\frac{\beta}{\beta_1 \mid \cdots \mid \beta_n},$$

with a sequence

$$\frac{\beta_1'}{\beta_1 \mid \beta_2'}$$

$$\frac{\beta_2'}{\beta_2 \mid \beta_3'}$$

$$\ddots$$

$$\frac{\beta_{n-1}'}{\beta_{n-1} \mid \beta_n}$$

of applications of the second of the rules given above, thus repeatedly extending the corresponding branch to the right as indicated above.

Thus we obtain a tree $\mathcal{T}''$ with the same number of branches as $\mathcal{T}$. The nodes of $\mathcal{T}''$ are marked with formulas of the language of TC_m. By translating these formulas to the language of C we obtain the wanted refutation $\mathcal{T}'$.

q.e.d.

Whereas the number of branches is the same for the two trees, the number of nodes may differ considerably as the following example of propositional logic shows. Let $S = \{\neg A_n, A_1 \wedge \cdots \wedge A_n\}$. Then the shortest refutation of S in TC_m is the non-branching tree consisting of three nodes marked $\neg A_n$, $A_1 \wedge \cdots \wedge A_n$ and A_n. The same is true in TC_{ac} and in TC_{iac} but not in TC where the shortest refutation has $n+1$ nodes.

For the calculi TC_{ac} and TC_{iac} all rules except the β-rule directly correspond to the respective rules of TC_m. For each β-formula β of TC_m and for $n \stackrel{\text{def}}{=} n[\beta]$ the number of new nodes introduced by application of the β-rule to β in TC_m equals n. In the transform of the tree to any of the other three calculi the number of new nodes introduced is $2n - 2$. So the total number of nodes of the tree increases at most by the factor 2 by the transformation from TC_m to TC_{ac} or to TC_{iac}.

For the transformation from TC_m to TC, no bound can be given for the increase of the number of nodes. But the worst case increase of the sum of the lengths of formulas marking the nodes, and also of the number of occurrences of literals in the tree, is exactly quadratic. It cannot be more than quadratic since the number of nodes introduced into $\mathcal{T}'$ for each application of an α-rule in $\mathcal{T}$ is at most equal to the number of literals in $\mathcal{T}$. The example above shows that the increase may, in fact, be as bad as quadratic.

From the facts stated so far, the number of branches of a refutation in a tableau calculus seems to be a more natural measure for the complexity of treating a problem with the method of tableaux than the number of nodes is. The number of branches is the number of cases needed to consider in a case analysis of the given problem whereas the number of nodes of the tableau depends very much upon how the mechanism of breaking up alpha formulas in the particular calculus is defined. Moreover, the number of nodes of a tableau for S can be bounded by the product of the number of branches of the tableau and of the length of the longest formula of S.

So we shall now consider the number of branches of tableaux. We shall prove that the reverse of Proposition 2.6.1 does not hold. The number of branches may increase when going from TC_{ac} or from TC_{iac} to TC or to TC_m. We first need a lemma for this. Since we are only interested in finite sets S of formulas here and only in the number of branches, we shall assume henceforth, without loss of generality, that all tableaux that we consider, are constructed by first introducing all elements of S and then only applying the α-, β-, γ- and δ-rules.

Lemma 2.6.1

Let S be a set of (possibly nested) disjunctions of literals of propositional logic and let $\mathcal{T}$ be a refutation of S in the calculus TC. Let κ be a node of $\mathcal{T}$ such that all elements of S are introduced somewhere on the path $\mathcal{P}$ from the root of $\mathcal{T}$ to the node κ. Let P be the set of formulas marking nodes on the path $\mathcal{P}$. Assume that there is a β-formula $\beta \in P$ and a literal $L \in P$ such that β_1 and F are complementary literals or β_2 and L are complementary literals. Further assume that $P \setminus \{\beta\}$ is satisfiable. Then there is a refutation $\mathcal{T}'$ of S in TC such that

1. *$\mathcal{T}'$ differs from $\mathcal{T}$ only in the subtree belonging to κ.*
2. *In $\mathcal{T}'$ the node κ has exactly two successors, and these are marked β_1 and β_2 (and obtained from β by application of the β-rule).*
3. *The number of branches of $\mathcal{T}'$ is less than or equal to the number of branches of $\mathcal{T}$.*

Proof: Without loss of generality we may assume that β_1 is complementary to L. Let $\mathcal{T}_1$ be the subtree of $\mathcal{T}$ belonging to the node κ. Since S contains only disjunctions of literals, the formula β can contribute to the closure of $\mathcal{T}$ only by application of the β-rule to β. Since $P \setminus \{\beta\}$ is satisfiable and since S contains only disjunctions of literals, there must be an application of the β-rule to β in the subtree $\mathcal{T}_1$ of $\mathcal{T}$. Now let $\mathcal{T}'$ be obtained from $\mathcal{T}$ by inserting an application of the β-rule to β at the node κ, thus producing a closed branch ending with β_1, and then removing all applications of the β-rule to β from the subtree $\mathcal{T}_1$. The tree $\mathcal{T}'$ is still a refutation of S. Altogether, we have added one new branch and removed one or more old branches. So the number of branches of $\mathcal{T}'$ is less than or equal to the number of branches of $\mathcal{T}$.

q.e.d.

If we want to find a refutation of a set of disjunctions of literals in TC which has a minimal number of branches then this lemma tells us that a safe rule to use is the following restriction of the β-rule which we shall call the rule of

UNIT REDUCTION $\qquad \dfrac{\beta}{\beta_1 \mid \beta_2}$

if deletion of β would result in a satisfiable branch, and after application of this rule the branch belonging to β_1 or the branch belonging to β_2 is closed.

This rule can be strengthened by defining the set $F[B]$ of non-used formulas of a branch B to be the set of all formulas F marking nodes of B such that no application of the β-rule to F occurs in the branch B. Then a branch B containing β can be extended according to the rule of unit reduction by appending two new successors marked β_1 and β_2 to the leaf of B, provided that $F[B] \setminus \{\beta\}$ is satisfiable, and some node of B is marked with a literal complementary to β_1 or to β_2.

Proposition 2.6.2

There is a set S of formulas of TC_m and a positive integer n such that there are refutations of S in TC_{ac} and in TC_{iac} with exactly n branches, but there do not exist any refutations of S in TC or in TC_m with at most n branches.

Proof: Consider the following set S of formulas of TC_m.

$$
\begin{aligned}
S = \{\ & A_1, \\
& B_1, \\
& A_2 \vee B_2, \\
& A_3 \vee B_2, \\
& A_2 \vee B_3, \\
& A_3 \vee B_3, \\
& \neg A_2 \vee \neg A_1 \vee \neg A_3, \\
& \neg B_2 \vee \neg B_1 \vee \neg B_3 \quad \}.
\end{aligned}
$$

A refutation of S in the calculus TC_{ac} and in TC_{iac} with exactly 10 branches is shown in Figure 2.5. In this figure, the application of the β-rule to the formula $\neg B_2 \vee \neg B_1 \vee \neg B_3$ yielding the formulas $\neg B_1$ and $\neg B_2 \vee \neg B_3$, and the subsequent application of the β-rule to the formula $\neg B_2 \vee \neg B_3$, have been merged for brevity, as in the proof of Proposition 2.6.1, without changing the number of branches of the tree. Of course, the full derivation tree is a binary tree.

If we try to obtain a refutation of S in TC then we have the choice to which element of S we want to apply the β-rule in the first step. If we apply it to the formula $\neg A_2 \vee \neg A_1 \vee \neg A_3$ then we get two successor nodes marked with $\neg A_2$ and $\neg A_1 \vee \neg A_3$, respectively. For the path to $\neg A_2$ we can successively apply unit reduction with the literal $\neg A_2$ and the formula $A_2 \vee B_2$, with $\neg A_2$ and $A_2 \vee B_3$, with B_2 and $\neg B_2 \vee \neg B_1 \vee \neg B_3$, with B_1 and $\neg B_1 \vee \neg B_3$, and finally with B_3 and $\neg B_3$. These are five branches. Similarly, for the path going through $\neg A_1 \vee \neg A_3$ we can repeatedly apply unit reduction and we obtain six more branches. So altogether we have 11 branches. If the first application of the β-rule is done to the formula $\neg B_2 \vee \neg B_1 \vee \neg B_3$ we obtain exactly the same result. If the first chosen β-formula of S is one of the four two-element disjunctions we need even 14 branches to close the tableau. So there is no refutation in TC which has 10 or less branches. From Proposition 2.6.1 it follows that there exists no refutation in TC_m of S which has at most 10 branches.

q.e.d.

2.6.3 The Tableau Calculus with Unification, TC_u

A feature that is essential for the efficiency of almost every theorem proving method is unification. This applies also to the tableau calculus. For simplicity and to obtain better comparability with other methods of automated theorem proving we consider clausal form logic. As has already been mentioned this does not mean a serious restriction, even not in terms of efficiency. So we assume that S is a set of clauses, i.e., of universal closures of disjunctions of literals. The only applicable rules except for the introduction of clauses of S, are the γ-rule and the β-rule. Moreover, an application of the γ-rule is useful only if the γ-rule or the β-rule is applied to the result afterwards (except in the case of a unit clause where the application of the β-rule is void). So we can restrict application by requiring that any application of a γ-rule must be followed immediately by an application of the γ-rule or of the β-rule to the result. In fact, we shall combine the β-rule and the γ-rule in one rule.

Definition 2.6.5
Let $\mathcal{T}$ be a tree whose nodes are marked with formulas, and let σ be a substitution of variables. By $\sigma(\mathcal{T})$ we denote the tree obtained from $\mathcal{T}$ by replacing every free occurrence of every variable x in every formula marking a node of $\mathcal{T}$ by the term $\sigma(x)$.

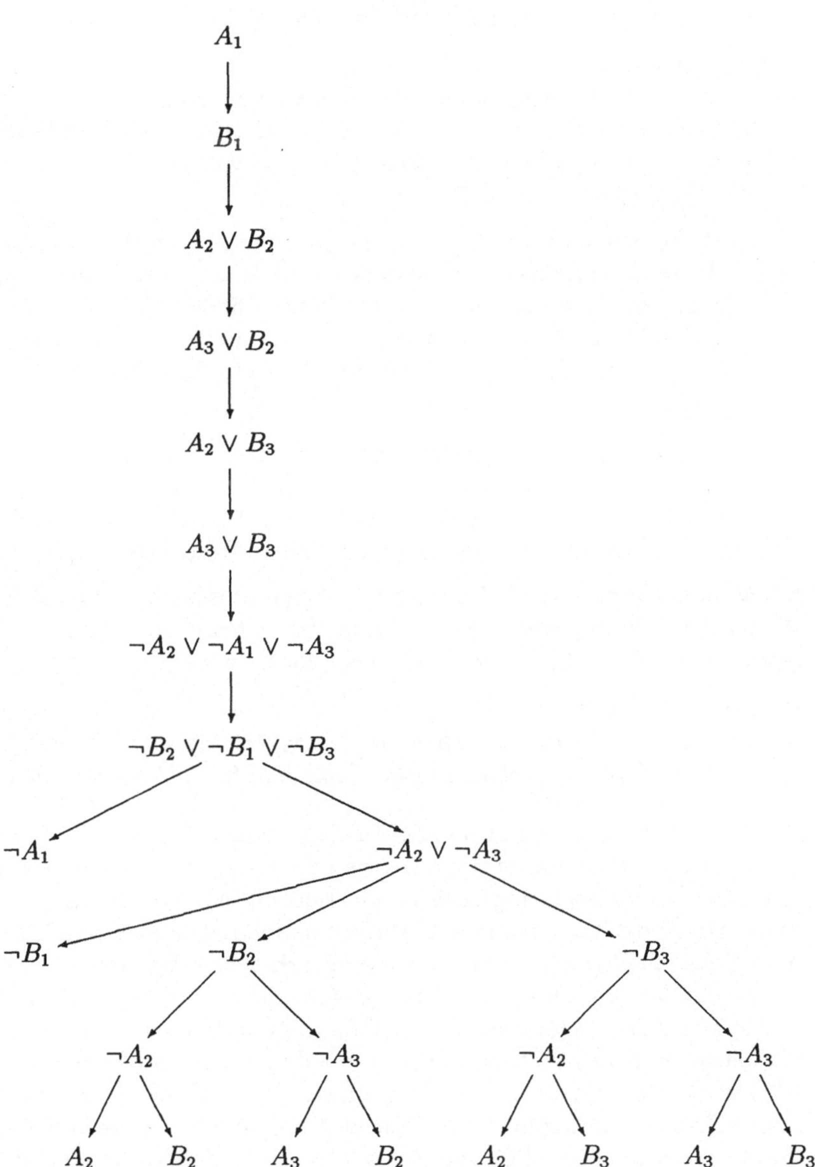

Figure 2.5: A refutation in TC_{ac}

If $\mathcal{T}$ is a tableau for a set S of closed formulas in one of the calculi TC, TC_{ac}, TC_{iac} or TC_m, and if σ is a substitution, then $\sigma(\mathcal{T})$ is also a tableau for S in the same calculus. Moreover, if $\mathcal{T}$ is closed then $\sigma(\mathcal{T})$ is also closed.

Definition 2.6.6
The tableau calculus with unification for formulas in clausal form, TC_u is defined by the following concepts and rules. An *initial tableau* for a set S of clauses is a unary tree (a chain) where each node is marked with an element of S. Now let $\mathcal{T}$ be a tree and B a branch of $\mathcal{T}$.

1. Let there be a clause $\forall x_1 \ldots \forall x_k(L_1 \vee \cdots \vee L_n)$ marking some node of B. Let π be some permutation of variables such that the variables $\pi(x_1), \ldots, \pi(x_k)$ do not occur in any formula marking a node of B. Let $\mathcal{T}'$ be the tree obtained from $\mathcal{T}$ by appending n successor nodes to B which are marked with $\pi(L_1), \ldots, \pi(L_n)$. Then we say that $\mathcal{T}'$ is obtained from $\mathcal{T}$ by *application of the γ-β-rule.*

2. Let K and L be two literals marking nodes of B. We assume that L marks the leaf of the branch B. Furthermore assume that K and L are complementary, i.e. , unifiable with opposite signs. Let σ be the most general unifier of K and L. Then we say that $\sigma(\mathcal{T})$ is obtained from $\mathcal{T}$ by *closing the branch B.*

A *tableau* in the calculus TC_u for a set S of clauses is a tree obtained from an initial tableau for S by repeated application of the γ-β-rule and closure of branches. A *refutation* of S in TC_u is a closed tableau for S in TC_u.

Proposition 2.6.3
Let S be a set of clauses. Then, for every refutation of S in TC_u there is a refutation of S in TC_m which has the same number of branches, and vice versa.

Proof: Let $\mathcal{T}$ be a refutation of S in TC_u. Then $\mathcal{T}$ is obtained from an initial tableau by a succession of applications of rules of TC_u giving a finite sequence of tableaux. By replacing each application of the γ-β-rule of TC_u to a clause $\forall x_1 \ldots \forall x_k D$ by k applications of the γ-rule followed by an application of the β-rule of the calculus TC_m we obtain a sequence of tableaux for S in TC_m. The last of these tableaux is closed and has the same number of branches as $\mathcal{T}$.

Now let $\mathcal{T}$ be a refutation of S in TC_m. Similarly as above we obtain a sequence of tableaux of S in TC_u from the corresponding sequence of tableaux of S in TC_m. We just ignore applications of the γ-rule and replace every application of the β-rule of TC_m by an application of the γ-β-rule of TC_u. Here the only problem is that in the γ-β-rule of TC_u the terms introduced have to be variables whereas in TC_m arbitrary terms are allowed. So a tableau $\mathcal{T}_u$ of our sequence of tableaux constructed for TC_u does not directly mirror the corresponding tableau $\mathcal{T}_m$ in TC_m. Rather, there is a substitution σ such that $\mathcal{T}_m = \sigma(\mathcal{T}_u)$. In order to obtain a closed tableau we add the application of closure of a branch in TC_u each time a branch closes in the corresponding tableau of TC_m.

q.e.d.

2.7 The Method of Tableaux and the Connection Method

In this section we give a comparison of the tableau calculus and the connection method. We show that the simplest version of the connection method without factorization can be directly simulated in the tableau calculus and vice versa, and how factorization can be incorporated in the tableau calculus.

Proposition 2.7.1
Let S be a set of clauses. Let n be a positive integer. Then, for each connection derivation of S which has exactly $n-1$ extension steps and no factorization steps there is a refutation of S in TC_u with exactly n applications of the γ-β-rule, and vice versa.

Proof: Let M be the matrix corresponding to S. Let $(S_0, \ldots, S_r)$ be a connection derivation of M. To each structured matrix $S_s = (M, p, D, \sigma)$ corresponds a tableau of TC_u as follows. If $S_0 = (M, (), (c))$ then the tableau corresponding to S_0 is the tableau obtained from the initial tableau by application of the β-γ-rule to c. If S_{s+1} is obtained from S_s by an extension step as in Definition 1.3.4 then (using the notations of Definition 1.3.4) the tableau corresponding to S_{s+1} is obtained from the tableau corresponding to S_s by application of the β-γ-rule to the clause c and the branch $(L_1, \ldots, L_{k+1})$ and subsequently closing all extensions of this branch that do not pass through a literal of e. If S_{s+1} is obtained from S_s by a truncation step then the tableau corresponding to S_{s+1} is identical to the tableau corresponding to S_s. The branches of the tableau thus constructed, correspond exactly to the paths occurring in a connection proof. The closed branches of the tableau correspond exactly to the paths that have already been checked for complementarity in the connection proof.

Now let there be a refutation $\mathcal{T}$ of S in TC_u. As we have just seen, the connection method can be mirrored in the tableau calculus, generating a tableau. The connection method traverses the nodes of the tableau it generates in a depth-first search manner. Obviously, the tableau $\mathcal{T}$ can be generated by a depth-first search sequence of applications of the γ-β-rule, interspersed with closures of branches each time a closed branch of $\mathcal{T}$ is encountered. The sequence of tableaux thus generated can be transformed to a connection proof corresponding to it in the same way as indicated above. We shall not give the formal proof here which is rather tedious but straight forward.

q.e.d.

The factorization rule of the connection method can also be directly incorporated in a tableau calculus by adding a *factorization rule* as follows. If $A = (K_1, \ldots, K_k, K)$ and $B = (K_1, \ldots, K_k, L_1, \ldots, L_l)$ are branches of the tableau, and K and L_l have identical sign and are unifiable, then the most general unifier of K and L_l may be applied to the tableau and, at the same time, the branch B

may be marked. A tableau for a formula is then said to be a *refutation* of that formula if each of its branches is closed or marked.

Proposition 2.7.2
The tableau calculus cannot p-simulate the tableau calculus with factorization.

Proof: Consider the following example of propositional logic taken from Cook and Reckhow's paper [CR74]. Let E denote the set of signs, $E = \{+, -\}$. For each natural number n and each sequence $(\epsilon_1, \ldots, \epsilon_n)$ of signs + and − let $P[\epsilon_1 \ldots \epsilon_n]$ be a propositional variable, and assume all of them are distinct. If P is a propositional variable then let $+P$ denote P and let $-P$ denote $\neg P$. Now, for each $n \in \mathbf{N}$, let

$$T_n \stackrel{\text{def}}{=} \{\epsilon_1 P[\,] \vee \epsilon_2 P[\epsilon_1] \vee \epsilon_3 P[\epsilon_1 \epsilon_2] \vee \cdots \vee \epsilon_n P[\epsilon_1 \ldots \epsilon_{n-1}] \mid \epsilon_1, \ldots, \epsilon_n \in E\}.$$

where $\vee$ is understood to be left associative. Then T_n is an unsatisfiable set of 2^n clauses, and it is known (see [CR74]) that there is a positive real number c such that every closed tableau for T_n has at least $2^{2^{cn}}$ nodes (whereas there is a resolution refutation for this set of clauses in $2^n - 1$ steps). We shall give a refutation of T_n in the tableau calculus with factorization, i.e., a tableau for T_n in which every branch is closed or marked, which has exactly $5 \cdot 2^n - 8$ nodes. So the number of nodes is linear in the number of clauses.

Let us denote by $F[\epsilon_1 \ldots \epsilon_n]$ the formula

$$\epsilon_1 P[\,] \vee \epsilon_2 P[\epsilon_1] \vee \epsilon_3 P[\epsilon_1 \epsilon_2] \vee \cdots \vee \epsilon_n P[\epsilon_1 \ldots \epsilon_{n-1}].$$

Then

$$F[\epsilon_1 \ldots \epsilon_{n+1}] \equiv F[\epsilon_1 \ldots \epsilon_n] \vee \epsilon_{n+1} P[\epsilon_1 \ldots \epsilon_n].$$

Now, the construction of the tableau starts with the introduction of all elements of T_n. Next, choose some element $F[\epsilon_1 \ldots \epsilon_n]$ of T_n and apply to it the β-rule to produce two branches, the first one B_1 with the leaf $F[\epsilon_1 \ldots \epsilon_{n-1}]$ and the second one B_2 with the leaf $\epsilon_n P[\epsilon_1 \ldots \epsilon_{n-1}]$. In B_2 we again apply the β-rule, now to $F[\epsilon_1 \ldots \epsilon_{n-1} \bar{\epsilon}_n]$ where $\bar{\epsilon}_n$ is the sign opposite to ϵ_n. So we get two subbranches, the first one B_{21} with the leaf $F[\epsilon_1 \ldots \epsilon_{n-1}]$, and the other one B_{22} with the leaf $\bar{\epsilon}_n P[\epsilon_1 \ldots \epsilon_{n-1}]$. Now, the branch B_{22} is closed and the branch B_{21} can be marked for factorization with the branch B_1. So we are left with just one branch B_1 that is neither closed nor marked, and the leaf of this branch is $F[\epsilon_1 \ldots \epsilon_{n-1}]$. Now, we can do this with every formula of T_n. The resulting branch contains all formulas of T_{n-1}. In a similar way we generated the formulas of T_{n-2} and so on, down to T_1 where the branch closes. For each of the 2^j formulas of T_j where $1 \leq j < n$, we have produced 4 new nodes in our tableau, and, moreover, we have the 2^n formulas of T_n. So the number of nodes of the tableau is $\sum_{j=1}^{n-1} 4 \cdot 2^j + 2^n$ which is $5 \cdot 2^n - 8$.

q.e.d.

Chapter 3

The Extension Rule in First Order Logic

In [Tse70], Tseitin introduced a structure-preserving transformation to clausal form for propositional logic, and the extension rule which is a deduction rule for clausal form formulas in propositional logic. He remarks that, using his transformation to clausal form, every derivation of a formula in the sequent calculus can be transformed in polynomial time to a refutation of the corresponding clausal form with the resolution rule and the extension rule. Moreover, if the derivation in the sequent calculus is cut-free then the corresponding refutation is a resolution refutation (without use of the extension rule). For a more detailed exposition of Tseitin's ideas and results see Reckhow's PhD thesis [Rec76].

In this chapter we generalize Tseitin's idea to first order logic and give a formal definition of the construction. Our generalization to first order logic of Tseitin's transformation is the transformation to definitional form described in Section 2.2. In this chapter, we generalize Tseitin's extension rule to first order logic. We combine the transformation to definitional form, the resolution calculus, and the extension rule to the calculus of *extended definitional resolution*. Then we prove analogous results to his for first order logic. We also consider derivations in extended definitional resolution in which the resolution part is restricted to tree resolution. Similar results on the p-simulatability of the natural deduction and Frege-Hilbert calculi in extended definitional resolution will then follow from Gerhard Gentzens simulation results in [Gen35].

The main results of this chapter are the following. Extended definitional tree resolution can p-simulate tree derivations of the sequent calculus, of the natural deduction calculus, and of the Frege-Hilbert calculus. Definitional tree resolution (without the extension rule) can p-simulate cut-free tree derivations in the sequent calculus (Theorem 3.6.1 in Section 3.6.1). Extended definitional resolution can p-simulate the sequent calculus, the natural deduction calculus, and the Frege-Hilbert calculus. Definitional resolution can p-simulate cut-free derivations in the sequent calculus (Theorem 3.6.2 in Section 3.6.2). Moreover, extended definitional resolution can p-simulate the sequent calculus augmented by the feature of definition, natural deduction augmented by the feature of definition, and the

Frege-Hilbert calculus augmented by the feature of definition (Theorem 3.8.1).

No effort has been made to make the estimates in this chapter optimal. It seems very likely that the degrees of the polynomials given as upper bounds for complexities in this chapter, can be reduced in some cases at least by 1 or 2. In an actual implementation, a further reduction would be obtained by structure sharing techniques, and it seems that a careful analysis of the simulations given in this chapter, and of the complexity of each single derivation step would reveal that these simulations are, in practise, much more efficient than suggested by the estimates proved in this chapter. Here we consider only sizes of derivations. We are not concerned with the time needed to check that a derivation is, in fact, a derivation, as long as we know that this check can be done in polynomial time. A much finer analysis of complexities would be needed to include this aspect.

3.1 The Extension Rule

The extension rule is a rule that allows to add the concept of definitions to a clausal form theorem proving calculus. Intuitively speaking, an extension step consists of a definition of a new predicate P in terms of one or two predicates Q (or Q and R). The predicates Q and R are assumed to be given predicates, or predicates that have been defined in an earlier extension step.

Definition 3.1.1
A set of clauses S' is obtained from a set of clauses S by *junctorial extension* if $S' = S \cup S_0$ where[1]

$$S_0 = \{\bar{J} \vee \bar{K} \vee \bar{L}, J \vee K, J \vee L\},$$

J has the form $P(x_1, \ldots, x_k)$, and the following three conditions hold.

1. P is a predicate symbol not occurring in S, possibly preceded by a negation sign.[2]

2. The variables $x_1, \ldots, x_k$ are pairwise distinct.

3. Every variable occurring in K or in L is in $\{x_1, \ldots, x_k\}$.

Note that the rule of junctorial extension is essentially Tseitin's extension rule, transferred to first order predicate logic.

Definition 3.1.2
A set of clauses S' is obtained from a set of clauses S by *quantorial extension* if $S' = S \cup S_0$ where

$$S_0 = \{J \vee \bar{K}, \bar{J} \vee L\},$$

[1]In this section we shall sometimes write clauses as disjunctions of literals (which are tacitly assumed to be universally quantified) to make them more readable. Note, that clauses really are sets of literals rather than disjunctions.

[2]The negation sign is not necessary, but it is convenient for us to obtain better correspondence to the definitional form.

J has the form $P(x_1, \ldots, x_k)$, and there is a variable x such that the following four conditions hold.

1. P is a predicate symbol not occurring in S, possibly preceded by a negation sign.

2. The variables $x_1, \ldots, x_k$ and the variable x are pairwise distinct.

3. Every variable occurring in K is in $\{x_1, \ldots, x_k, x\}$.

4. There is an n-ary function symbol f not occurring in S such that L is obtained from K by replacing all occurrences of x in K by $f(x_1, \ldots, x_k)$.

The function f is called the *Skolem function* of the quantorial extension step.

Definition 3.1.3
A set of clauses S' is obtained from a set of clauses S by *extension* if S' is obtained from S by junctorial extension or by quantorial extension.

Proposition 3.1.1
Let S be a set of clauses and let S' be obtained from S by extension. Then S' is satisfiable iff S is satisfiable. Moreover, if F is a formula in which neither the predicate symbol nor any Skolem function introduced in the extension, occurs, then it holds that $S' \models F$ iff $S \models F$.

Proof: Obviously, if S' is satisfiable then so is S. Now let S be satisfiable. Then there is an interpretation ι that satisfies S. In the case of a junctorial extension step let the interpretation ι' be defined as ι except that the predicate symbol P is interpreted so as to make the formula $J \leftrightarrow \neg K \vee \neg L$ true. Then ι' satisfies S'. In the case of a quantorial extension step let ι' be defined as ι except that P is interpreted so as to make the formula $J \leftrightarrow \exists x K$ true, and the function symbol f is interpreted so as to make $\exists x K \rightarrow L$ true. Then ι' again satisfies S'. The proof for the second assertion is quite analogous.

q.e.d.

Let us remark that the extension rule is not sound, i.e., if S' is obtained from S by extension then $S \models S'$ does not hold in general.

Definition 3.1.4
A *derivation* of a clause c from a set of clauses S in *extended resolution*, or an *ER-derivation*, is a finite sequence $(c_1, \ldots, c_n)$ of clauses such that the following three statements hold.

1. $c_n = c$.

2. For all $k = 1, \ldots, n$, at least one of the following five conditions holds.

   (a) $c_k \in S$.

(b) c_k is a factor of some clause c_j with $j < k$.

(c) c_k is a resolvent of two clauses c_i and c_j with $i, j < k$.

(d) $\{c_1, \ldots, c_{j+3}\}$ is obtained from $\{c_1, \ldots, c_j\}$ by a junctorial extension for some j with $j < k \leq j+3$.

(e) $\{c_1, \ldots, c_{j+2}\}$ is obtained from $\{c_1, \ldots, c_j\}$ by a quantorial extension for some j with $j < k \leq j+2$.

3. None of the predicate symbols newly introduced in the extension steps and none of the Skolem functions of the extension steps occur in the clause c.

A *refutation* of a set of clauses S in *extended resolution* is a derivation of the empty clause from S in extended resolution.

A clause is said to be *tautological* if it contains a literal L and its negation $\neg L$. For the resolution calculus the following soundness and completeness theorem holds.

Proposition 3.1.2
Let S be a set of clauses and let c be a clause. Then the following two statements hold.

1. *If there is a derivation of c from S in the resolution calculus then $S \models c$.*

2. *If $S \models c$, and c is not tautological, then there is a clause d subsuming the clause c such that there is a derivation of d from S in the resolution calculus.*

From Proposition 3.1.1 and from the soundness and completeness theorem for the resolution calculus follows immediately the following soundness and completeness theorem for extended resolution.

Proposition 3.1.3
Let S be a set of clauses and let c be a clause. Then the following two statements hold.

1. *If there is a derivation of c from S in extended resolution then $S \models c$.*

2. *If $S \models c$, and c is not tautological, then there is a clause d subsuming the clause c such that there is a derivation of d from S in extended resolution.*

In particular, a set of clauses is refutable in extended resolution iff it is unsatisfiable. Now, consider the set of clauses def_F introduced in Definition 2.2.1 for each formula F.

Definition 3.1.5
Let F be a formula of full first order logic. Then a *proof* or *derivation* of F *in extended definitional resolution* is a derivation of the unit clause $\{L_F\}$ from the clause set def_F in extended resolution. A *proof* or *derivation* of F in *definitional resolution* is a resolution derivation of $\{L_F\}$ from def_F.

Proposition 3.1.4
Let F be a formula of full first order logic. Then the following three statements are equivalent to each other.

1. *F is valid.*
2. *There exists a proof of F in definitional resolution.*
3. *There exists a proof of F in extended definitional resolution.*

Proof: It is known from the literature cited in Section 2.2 that F is valid if and only if $\mathrm{def}_F \models L_F$ holds. The proof is very similar to the proof of Proposition 3.1.1.

Now, if there exists a proof of F in definitional resolution or in extended definitional resolution, then there exists by definition a derivation of $\{L_F\}$ from def_F in resolution or in extended resolution, respectively. From the soundness theorems for resolution and for extended resolution (Proposition 3.1.3, part 1), it follows that $\mathrm{def}_F \models L_F$, and therefore F is valid.

If, on the other hand, F is valid, then $\mathrm{def}_F \models L_F$ holds. From the completeness theorem for resolution it follows that there is a resolution derivation $\mathcal{R}'$ of a clause c from def_F such that the clause c subsumes the clause $\{L_F\}$. But the clause $\{L_F\}$ is a unit clause of the form $P(x_1, \ldots, x_k)$ where $x_1, \ldots, x_k$ are pairwise distinct variables. So c must be a variant of $\{L_F\}$. Let $\mathcal{R}$ be the sequence of clauses obtained form $\mathcal{R}'$ by replacing its last clause c by $\{L_F\}$. Then $\mathcal{R}$ is a resolution derivation of L_F from def_F. So, $\mathcal{R}$ is a proof of F in definitional resolution. Of course, $\mathcal{R}$ is also a proof of F in extended definitional resolution.

q.e.d.

We shall prove in this chapter that extended definitional resolution can p-simulate the sequent calculus, and that definitional resolution can p-simulate the sequent calculus without the cut rule.

3.2 Complexities of Formulas and Derivations

In Section 3.6, we shall compare the complexities of extended definitional resolution and of the sequent calculus. In order to be able to do this we define a few complexity measures for these two calculi in this section.

Definition 3.2.1
The *structural complexity* of a formula is the sum of the number of occurrences of logical symbols $\neg$, $\vee$, $\wedge$, $\exists$, $\forall$, and of the number of occurrences of predicate symbols in the formula. The *degree* of a formula is the number of occurrences of logical symbols in it.

Definition 3.2.2
Let A be a finite or infinite alphabet. Let w be a word over A. Then the *size* $|w|$ of the word w is the number of occurrences of symbols in w.

We shall regard a derivation in resolution with subsumption as a word over an infinite[3] alphabet consisting of the variables, the function symbols, the predicate symbols, the auxiliary symbols '(','\)' and ',', and of the symbols $\neg$, $\vee$, and #. The *representation of a literal* L as a word is the literal L itself. A *representation of a clause* as a word is the disjunction of its literals. The empty clause is represented by the empty word.[4]

Instead of 'representation as a word' we shall also say '*word representation*'. Now, let # be a symbol that does not occur in any representation of a clause. By the *word representation of a sequence of clauses* $(c_1, \ldots, c_r)$ we mean the word $w_1\# \ldots \#w_r\#$ where $w_1, \ldots, w_r$ are word representations of $c_1, \ldots, c_r$, respectively. By a *word representation of a set of clauses* S we mean a word representation of a sequence $(c_1, \ldots, c_r)$ of clauses such that $\{c_1, \ldots, c_r\} = S$.

Definition 3.2.3
By a *derivation* of a set of clauses S_2 from a set of clauses S_1 by *resolution with subsumption* we mean a word representation of a sequence $(c_1, \ldots, c_n)$ of clauses such that the following conditions are fulfilled.

1. $S_2 \subseteq S_1 \cup \{c_1, \ldots, c_n\}$.
2. For each $k = 1, \ldots, n$, there is a clause $c \in S_1 \cup \{c_1, \ldots, c_{k-1}\}$ subsuming the clause c_k, or there are clauses $c, d \in S_1 \cup \{c_1, \ldots, c_{k-1}\}$ such that the clause c_k is subsumed by a resolvent of c and d.

We shall frequently identify clauses with representations of clauses and use the word 'clause' to denote both of these concepts if it is clear from the context what is meant.

Lemma 3.2.1
If S_1, S_2, S_1', S_2' are sets of clauses and $S_1 \subseteq S_1'$ and $S_2 \supset S_2'$ then every derivation of S_2 from S_1 by resolution with subsumption is a derivation of S_2' from S_1' by resolution with subsumption.

Definition 3.2.4
Let Q be a derivation by resolution with subsumption. Then the *C-complexity* $|Q|_C$ of Q is the number of clauses in Q. The *L-complexity* $|Q|_L$ of Q is the number of literal occurrences[5] in Q. The *size* $|Q|$ of Q is its size as a word, as defined above.

[3]We use an infinite alphabet only for convenience. The infinitely many variables can be coded as words over a finite alphabet. Thus, every derivation of size n using our infinite alphabet, can be converted to a derivation of a size of the order $O(n \log n)$ using a finite alphabet.

[4]There are infinitely many representations for any non-empty clause, since in any representation, the order of the literals can be exchanged at will, and literals can occur repeatedly. But representations can always be normalized by demanding that no literal occurs twice, and then choosing the first representation in lexicographic order that obeys this condition.

[5]By 'literal occurrences' we mean here occurrences of literals as elements of clauses of Q. If A is an atomic formula and $\neg A$ occurs in a clause of Q then only the occurrence of $\neg A$ is counted, not the occurrence of the literal A as a suboccurrence of $\neg A$.

Lemma 3.2.2
If $\mathcal{Q}_1$ is a derivation of a clause set S_2 from a clause set S_1, and $\mathcal{Q}_2$ is a derivation of a clause set S_3 from the clause set S_2 by resolution with subsumption, then $\mathcal{D}_1\mathcal{D}_2$ is a derivation of S_3 from S_1 by resolution with subsumption. Moreover, $|\mathcal{Q}_1\mathcal{Q}_2|_C = |\mathcal{Q}_1|_C + |\mathcal{Q}_2|_C$, $|\mathcal{Q}_1\mathcal{Q}_2|_L = |\mathcal{Q}_1|_L + |\mathcal{Q}_2|_L$, and $|\mathcal{Q}_1\mathcal{Q}_2| = |\mathcal{Q}_1| + |\mathcal{Q}_2|$.

The concepts of a derivation, and of the C-complexity, the L-complexity, and the size of a derivation, are defined for extended resolution with subsumption in an analogous manner as for ordinary resolution with subsumption.

In the sequel, we shall frequently construct derivations by resolution with subsumption which we shall afterwards convert to resolution derivations. The following proposition states that such a conversion is possible. Remember Definition 2.3.5 of the concept of n-subsumption meaning a subsumption without merging.

Proposition 3.2.1
Let S be a set of clauses and let c be a clause. Let $\mathcal{Q}$ be a derivation of $\{c\}$ from S by resolution with subsumption. Let N be the size of $\mathcal{Q}$ and let n be the maximum of the cardinalities of clauses occurring in S or in $\mathcal{Q}$. Let M be the size of a word representation of S. Then there is a clause d n-subsuming the clause c, and there is a resolution derivation $\mathcal{R}$ of d from S such that $|\mathcal{R}| \leq 2(M+N)n^2$. If $\mathcal{Q}$ is a tree derivation then so is $\mathcal{R}$.

Proof: Let $\mathcal{C}$ be a word representation of a sequence built from the clauses of S. Let $\mathcal{F} \stackrel{\text{def}}{=} \mathcal{CQ}$. Then the size of $\mathcal{F}$ is $M+N$. The maximum of the cardinalities of clauses in $\mathcal{F}$ is n. Assume $\mathcal{F} = c_1\#\ldots\#c_r\#$. For $k = 1,\ldots,r$ we define a clause d_k and a (word representation of a) resolution derivation $\mathcal{R}_k$ of d_k from $S \cup \{d_1,\ldots,d_{k-1}\}$ such that d_k n-subsumes c_k. The definition is as follows.

If $c_k \in S$ then let $d_k \stackrel{\text{def}}{=} c_k$ and $\mathcal{R}_k \stackrel{\text{def}}{=} c_k\#$.

If c_k is subsumed by c_i with $i < k$ then $\mathcal{R}_k \stackrel{\text{def}}{=} e_1\#\ldots\#e_q\#$ where e_1 is a factor of c_i, each e_{p+1} is a factor of e_p, and e_q subsumes c_k. The clause d_k is defined to be the clause e_q. Here we assume that there is one-to-one correspondence between the literals of d_i and the literals of c_i. Each literal of d_i subsumes its corresponding literal in c_i. In particular, the cardinalities of c_i and of d_i are identical to each other. The factorization steps are to be done in such a way that a similar correspondence is established between c_k and d_k.

If c_k is subsumed by a resolvent of c_i and c_j where $i,j < k$ then let e_1 be the corresponding resolvent of d_i and d_j, and let $\mathcal{R}_k \stackrel{\text{def}}{=} e_1\#\ldots\#e_q\#$ and $d_k \stackrel{\text{def}}{=} e_q$ where each e_{p+1} is a factor of e_p and a similar correspondence holds between c_k and d_k as above.

Then the cardinality of the first clause in $\mathcal{R}_k$ is less than or equal to $2n-2$. The cardinality of each subsequent clause in $\mathcal{R}_k$ is smaller than the cardinality of its predecessor. So the number of literals in $\mathcal{R}_k$ is less than or equal to $2n^2$. Each literal in $\mathcal{R}_k$ subsumes a literal in c_k, and therefore its size is at most equal to the size of that literal of c_k. So, for each literal L in $\mathcal{R}_k$, the size of $L\#$ is less than or equal to the size of $c_k\#$. So $|\mathcal{R}_k| \leq 2 \cdot |c_k\#|n^2$. Now, let $\mathcal{R} \stackrel{\text{def}}{=} \mathcal{R}_1 \ldots \mathcal{R}_r$. Then it follows from Lemma 3.2.2 that $|\mathcal{R}| \leq 2(M+N)n^2$.

q.e.d.

Proposition 3.2.2
Let S be a set of clauses not containing any of the predicate symbols P_G or function symbols f_G introduced in Section 2.2. Let F be a formula and def_F *its definition. Let n be the structural complexity of F. Then the clause set $S \cup \text{def}_F$ can be obtained from the clause set S by a derivation $\mathcal{E}$ consisting of n extension steps. Moreover, the following inequalities hold.*

$$\begin{aligned} |\mathcal{E}|_C &\leq 3n \\ |\mathcal{E}|_L &\leq 7n \\ |\mathcal{E}| &\leq (5 \cdot |F| + 8)n \end{aligned}$$

The size of any word representation of $|\text{def}_F|$ equals $|\mathcal{E}|$. All clauses of def_F *have cardinalities less than or equal to* 3.

Proof: Each of the sets of clauses C_G, where G is a subformula of F, can be appended to S in one extension step. There are at most 3 clauses and 7 literals in C_G. In the notation used in the definition of the clause set C_G in Section 2.2, the following inequalities hold as is easily verified. $|L_G| \leq |G| \leq |F|$, and $|L_H|+|L_I| \leq |H| + |I| \leq |F| - 1$ and $|L| \leq 2 \cdot |G| - 6 < 2 \cdot |F|$, in the respective cases. In each case, the sum of the sizes of all occurrences of L_G, L_H, L_I, and L, in the clause set C_G is at most $5 \cdot |F| - 2$. There are at most four occurrences of $\vee$, three occurrences of $\neg$, and three occurrences of $\#$ in a representation of the clause set C_G as a word. So the size of a representation of C_G is at most $5 \cdot |F| + 8$. The number of sets C_G needed for the derivation is at most equal to the structural complexity n of F.

q.e.d.

3.3 Occurrences in the Sequent Calculus

In this section we make a few definitions concerning the sequent calculus which we shall need in the simulation of the sequent calculus by extended definitional resolution in Section 3.6.

We shall assume as in Section 1.6 that the set of variables is partitioned in the set of free variables and the set of bound variables. Moreover, we restrict the syntax of a formula by disallowing formulas that have bound occurrences of free variables. Furthermore, we disallow formulas that have subformulas in which a variable has free and bound occurrences. If a formula, in addition, has no free occurrences of bound variables, then we shall call it a *wellformed formula*. So each subformula of a formula is again a formula, but not each subformula of a wellformed formula is a wellformed formula. In derivations in the sequent calculus we allow only wellformed formulas. By renaming of variables, any derivation in the sequent calculus that does not obey this condition can be transformed to one that does obey it, without increasing its size.

In the sequel, we shall frequently speak about occurrences of formulas in sequents and in derivations. We shall say that a formula occurs as a *member* of a sequent $A_1, \dots, A_m \Longrightarrow B_1, \dots, B_n$ if it is one of the formulas $A_1, \dots, A_m$, $B_1, \dots, B_n$. We shall say that a formula *occurs as a subformula* in a sequent if it is a subformula of a member of the sequent. We shall say a formula occurs as a subformula in a derivation if it occurs as a subformula in some sequent of the derivation. We shall assume that derivations are given as strings of the form $\#\mathcal{S}_1\#\dots\#\mathcal{S}_r\#$ over a suitable (infinite alphabet), where $\mathcal{S}_1, \dots, \mathcal{S}_r$ are sequents and the character '#' does not occur in any sequent.

Definition 3.3.1

An *occurrence* of a nonempty string u *as substring of a string* w is a pair (j, k) of positive integers such that u is the substring of w which starts at the j-th character position of w and ends at the $(k-1)$-th character position of w. So the size of u is $k-j$. If we speak of an *occurrence of a formula* in a derivation or in a sequent of a derivation, we always mean its occurrence as a substring of the whole derivation. We never mean its occurrence as a substring of the sequent. By an *occurrence of a sequent in* a derivation $\mathcal{D}$ we mean an occurrence (j, k) of a sequent as a substring of $\mathcal{D}$ such that the $(j-1)$-th and the k-th character positions in $\mathcal{D}$ are occupied with '#'.

When we speak of occurrences we shall always assume (unless otherwise stated) that we are given a fixed derivation, and all occurrences are understood to be relative to this fixed derivation, i.e., they will be occurrences as substrings of the derivation. We shall not always explicitly mention the derivation.

We shall use small Greek letters to denote occurrences of formulas, namely α, β, γ to denote an occurrence of a formula as a member of a sequent in a derivation, and η, θ to denote an occurrence as a subformula. Occurrences of sequents in derivations are denoted by the small Greek letters χ and ψ. For any occurrence of a formula or of a sequent denoted by some small Greek letter, we denote by this letter with a tilde above it, the corresponding formula or sequent itself. For example, if α is an occurrence of the formula $\neg\forall x p(a, x)$ in a derivation then $\tilde{\alpha}$ is the formula $\neg\forall x p(a, x)$ itself. If β is another occurrence of the same formula then $\tilde{\alpha} \equiv \tilde{\beta}$, but $\alpha \neq \beta$.

If $\eta_1, \dots, \eta_n$ are occurrences such that there are positive integers $j_0, \dots, j_n$ such that $\eta_r = (j_{r-1}, j_r)$ for $r = 1, \dots, n$, then the occurrences $\eta_1, \dots, \eta_n$ are said to be *adjacent* occurrences. In this case the *concatenation* $\eta_1 \dots \eta_n$ of $\eta_1, \dots, \eta_n$ is defined to be the occurrence (j_0, j_n). In denoting a concatenation of occurrences we allow to write $\tilde{\eta}_r$ instead of η_r for some of the r's. For example, if η and θ are occurrences of formulas and ζ is an occurrence of $\wedge$ such that η, ζ and θ are adjacent, then we shall write $\eta \wedge \theta$ instead of $\eta\zeta\theta$.

Definition 3.3.2
Let λ be an occurrence of a formula, of a sequence of formulas, or of a sequent. Let k be a positive integer and let $\eta_1, \ldots, \eta_k$ be occurrences such that one of the following conditions holds.

1. $k = 1$, and either $\lambda = \neg\eta_1$ or $\lambda = \forall x \eta_1$ or $\lambda = \exists x \eta_1$ where x is a bound variable.

2. $k = 2$, and either $\lambda = \eta_1 \wedge \eta_2$ or $\lambda = \eta_1 \vee \eta_2$.

3. λ is an occurrence of a sequence of k formulas, $\lambda = \eta_1, \ldots, \eta_k$.

4. λ is an occurrence of a sequent, $\lambda = \eta_1, \ldots, \eta_j \Longrightarrow \eta_{j+1}, \ldots, \eta_k$.

Let $1 \leq j \leq k$. Then η_j is said to be the j-th *immediate suboccurrence* of λ, and $\tilde{\eta}_j$ is said to be the j-th *immediate subformula* of $\tilde{\lambda}$. In the last case, $\eta_1, \ldots, \eta_j$ are said to be occurrences in the *antecedent* of λ, and $\eta_{j+1}, \ldots, \eta_k$ are said to be occurrences in the *succedent* of λ. An occurrence θ is said to be a *suboccurrence* of an occurrence λ if there is a nonnegative integer n and a sequence $(\lambda_0, \ldots, \lambda_n)$ such that $\lambda_0 = \lambda$ and $\lambda_n = \theta$ and, for all $j = 1, \ldots, n$, the occurrence λ_j is an immediate suboccurrence of λ_{j-1}.

In the definition of the rules of the sequent calculus in Section 1.6, Latin letters 'A', 'B', 'D' and 'E' have been used to denote formulas, Greek letters 'Γ', 'Δ', 'Θ' and 'Λ' have been used to denote finite sequences of formulas, and the small Latin letters 'a', 'x' and 't' have been used to denote a free variable, a bound variable, and a term, respectively. By the word '*instantiation*' we mean the process of passing from expressions possibly containing some of these letters to the expressions they denote. By an *instance* of a rule of the sequent calculus we mean the result of consistently replacing in the rule the letters by formulas, sequences of formulas and terms, subject to the restrictions stated in Section 1.6. By the *instance* of some particular letter we mean the formula, sequence of formulas, or term, it is replaced with.

Definition 3.3.3
An *application* of a rule R in a derivation $\mathcal{D}$ is an $(n+1)$-tuple $(\chi_1, \ldots, \chi_n, \psi)$ of occurrences of sequents in $\mathcal{D}$ such that $\tilde{\chi}_1, \ldots, \tilde{\chi}_n$ are the premises of an instance R' of R and $\tilde{\psi}$ is the conclusion of R'. The occurrences $\chi_1, \ldots, \chi_n$ are called the *premises* of the rule application, and the occurrence ψ is called the *conclusion* of the rule application.

Let R' be an application of a rule R, and let S be a premise or the conclusion of R. Assume the letter 'A' occurs in S. Let χ be the premise of R', or the conclusion of R', respectively, that corresponds to S. Let η be the corresponding occurrence of the instance of 'A', so that η is a suboccurrence of χ. Then η is said to be an A-occurrence in R'. Similarly we define B-, D-, E-, Γ-, Δ-, Θ-, Λ-, F-, $F\{x\backslash a\}$-, and $F\{x\backslash t\}$-occurrences. In the conclusion of a rule application, there is always at most one A-occurrence, at most one B-occurrence, etc.

If η is an occurrence of a formula as a member or as a subformula in the premise of a rule application then we define the *immediate subordinate* of η which is essentially the corresponding occurrence of the same formula in the conclusion of the rule application. Only in the case of the formula $F\{x\backslash a\}$ or $F\{x\backslash t\}$ in a $\forall$- or $\exists$-inference it is not an occurrence of this formula, but an occurrence of the formula F, and in the case of a cut formula there is no immediate subordinate.

Definition 3.3.4
Let χ be a premise of a rule application R in a derivation $\mathcal{D}$, and let ψ be its conclusion. Let η be an occurrence of a formula H as a subformula in χ. Then, if R is an application of the cut rule and η is an occurrence of H as a subformula of the cut formula of R, then η has *no immediate subordinate* in R. Otherwise the *immediate subordinate* of η in R is an occurrence of a formula as a subformula in ψ, and it is defined as follows.

1. If η is an A-, B-, D-, or E-occurrence in χ, then the immediate subordinate of η is the A-, B-, D-, or E-occurrence, respectively, in ψ.

2. If η is a '$F\{x\backslash a\}$'- or '$F\{x\backslash t\}$'-occurrence in χ then the immediate subordinate of η is the 'F'-occurrence in ψ.

3. If η is the n-th immediate suboccurrence of a Γ-, Δ-, Θ-, or Λ-occurrence in χ, then the immediate subordinate of η is the n-th immediate suboccurrence of the Γ-, Δ-, Θ-, or Λ-occurrence in ψ.

4. If η is the n-th immediate suboccurrence of θ, and θ has an immediate subordinate ζ, then the immediate subordinate of η is the n-th immediate suboccurrence of ζ.

If the immediate subordinate of an occurrence with respect to a rule application exists, then it is uniquely determined.

Definition 3.3.5
Let R be a rule application, and let α be an occurrence of a formula as a member of a premise of R. Let η be the immediate subordinate of α in R. Then the *immediate descendant* of α in R is that formula occurrence β as a member of the conclusion of R such that η is a suboccurrence of β.

An occurrence of a formula as a member of a premise of a rule application has no immediate descendant if it is the occurrence of the cut formula, and it has exactly one immediate descendant otherwise.

Definition 3.3.6
Let R be an application of an operational inference rule of the sequent calculus. Then the *principal occurrence* of R is that immediate suboccurrence β of the

conclusion of R, for which there is an immediate suboccurrence α of one of the premises of R, such that β is an immediate descendant of α, and $\tilde{\alpha} \not\equiv \tilde{\beta}$. The formula $\tilde{\beta}$ is then said to be the *principal formula* of the rule application.

For every application of an operational inference rule there is one and only one principal occurrence, and one and only one principal formula. Applications of structural inference rules have no principal occurrences and no principal formulas.

Definition 3.3.7
Let $\mathcal{D}$ be a derivation, and let η and θ be occurrences of formulas in $\mathcal{D}$. Then θ is said to be an *immediate subordinate* or an *immediate descendant* of η if there is a rule application R in $\mathcal{D}$ such that θ is an immediate subordinate or an immediate descendant, respectively, of η in R. The relation '*... is a subordinate of...*' is the reflexive transitive closure of the relation '...is an immediate subordinate of...', and the relation '*... is a descendant of...*' is the reflexive transitive closure of the relation 'is an immediate descendant of'.

Definition 3.3.8
Let $\mathcal{D}$ be a derivation in the sequent calculus, and let ψ be the occurrence of the endsequent of $\mathcal{D}$. We say that an occurrence η of a formula as a subformula in $\mathcal{D}$ is a *terminal occurrence* if it is a suboccurrence of ψ, or a suboccurrence of an occurrence of a formula as the cut formula of some application of the cut rule in $\mathcal{D}$. A *terminal subordinate* of an occurrence η in $\mathcal{D}$ is a subordinate of η which is a terminal occurrence. A *terminal descendant* of an occurrence α in $\mathcal{D}$ is a descendant of α which is a terminal occurrence.

Note that, if η is an occurrence of a formula in a derivation, and if η has no immediate subordinate, then η is a terminal occurrence. In the case of a tree derivation (see Definition 3.3.10), the converse is also true.

Definition 3.3.9
Let $\mathcal{D}$ be a derivation in the sequent calculus, and let η be an occurrence of a formula as a subformula in $\mathcal{D}$. Then a *path* for η is a sequence $(\eta_0, \ldots, \eta_n)$ of occurrences such that the following conditions hold.

1. $\eta_0 = \eta$.

2. η_i is an immediate subordinate of η_{i-1} for $i = 1, \ldots, n$.

3. η_n is a terminal occurrence.

Lemma 3.3.1
For every occurrence of a formula as a subformula in a derivation, there is at least one path.

Proof: Assume the assertion were false. We know that any occurrence that has no immediate descendant, is a terminal occurrence. Therefore an induction argument shows that there is an infinite sequence of occurrences $(\eta_0, \eta_1, \ldots)$ such that $\eta_0 = \eta$ and, for all $i > 1$, the occurrence η_i is an immediate subordinate of η_{i-1}. But, if the occurrences of sequents in the derivation are $\chi_0, \ldots, \chi_r$ in their proper order, then again an induction argument shows that, for each $i \geq 0$, the occurrence η_i is a suboccurrence of χ_j for some $j \geq i$. This gives a contradiction for $j = r + 1$.

q.e.d.

Lemma 3.3.2
Let $\mathcal{D}$ be a derivation of a sequent $\mathcal{S}$ in the sequent calculus, and let ψ be the occurrence of $\mathcal{S}$ as the endsequent of $\mathcal{D}$. Then the following statements hold.

1. *Every occurrence of a formula in $\mathcal{D}$ as a subformula has a terminal subordinate in $\mathcal{D}$.*

2. *Every occurrence of a formula as a member of a sequent in $\mathcal{D}$ has a terminal descendant in $\mathcal{D}$.*

Proof: By the last lemma, there exists a path for the given occurrence. The last element of the path is then a terminal subordinate of the given occurrence.

q.e.d.

Definition 3.3.10
By a *tree derivation* in the sequent calculus we mean a derivation in which each occurrence of a sequent is used at most once as a premise of a rule. The *tree expansion* of an arbitrary derivation is the deduction tree corresponding to the given derivation.

The tree expansion of a tree derivation can be obtained in essentially linear time whereas for arbitrary derivations this is possible only in exponential time and space. We shall sometimes use the term 'tree expansion' of a derivation $\mathcal{D}$ also for a tree derivation having the same tree expansion as the derivation $\mathcal{D}$.

Lemma 3.3.3
Each occurrence of a formula in a tree derivation has at most one immediate subordinate and at most one immediate descendant.

Proof: Let η be an occurrence of a formula in the tree derivation. By definition of a tree derivation, there is only one rule application which has η as a premise. Furthermore, we know that, for each rule application which has η as a premise, there is at most one immediate subordinate.

q.e.d.

Lemma 3.3.4
In a tree derivation, for each occurrence of a formula as a subformula, there is one and only one path.

Proof: We have already proved the existence of a path. From the last lemma it follows by induction on n that, for each nonnegative integer n, and for any two paths for η whose lengths are at least n, the initial subsequences consisting of the first n elements of the two paths, respectively, are identical. Now, the assertion follows from the fact that terminal occurrences do not have immediate subordinates in tree derivations.

q.e.d.

Lemma 3.3.5
Every occurrence of a formula in a tree derivation has exactly one terminal subordinate. Every occurrence of a formula as a member of a sequent in a tree derivation has exactly one terminal descendant.

Proof: The existence is assured by Lemma 3.3.2. Let $\mathcal{D}$ be a tree derivation. Then no terminal occurrence in $\mathcal{D}$ has an immediate subordinate in $\mathcal{D}$. From this fact and from the last lemma the uniqueness follows by induction on the length of the derivation.

q.e.d.

3.4 Application of Substitutions to Formulas

In this section we introduce the concept of the application of a substitution to a formula. We shall investigate pairs of formulas F and G which can be made identical in form by application of substitutions, i.e.[6], $\sigma F \equiv \tau G$. We shall consider the atomic formulas L_F introduced in Section 2.2 for the transformation of a formula to clausal form. We shall prove that, if F and G are formulas and σ and τ are substitutions such that $\sigma F \equiv \tau G$, then there is a short derivation, by resolution with subsumption, of the clause $\neg\sigma L_F \vee \tau L_G$ from the clause set $\mathrm{def}_F \cup \mathrm{def}_G$.

Definition 3.4.1
Let F be a formula and σ a substitution such that the domain of σ is a subset of the set of bound variables, and such that, for each x in the domain of σ, the term σx does not contain any bound variables. Then we denote by σF the result of replacing all free occurrences of any bound variable x in F by σx.

The following lemma is obvious.

[6]If F and G are formulas then '$F \equiv G$' means that the formulas F and G are identical, viewed as strings of symbols.

Lemma 3.4.1
Let F be a formula and σ a substitution such that σF is defined. Then $|\sigma L_F| \leq |\sigma F|$.

Definition 3.4.2
Let F be a formula. Then a *good substitution* for F is a substitution σ such that for all variables x in the domain of σ the following conditions hold.

1. x is a bound variable.

2. No bound variables occur in σx.

3. x does not occur bound in F.

Note that the domain of a good substitution for F may contain bound variables which are free in F. Now, the following lemmata are immediate consequences of these definitions.

Lemma 3.4.2
If σ is a good substitution for a formula F then it is a good substitution for any subformula of F.

Lemma 3.4.3
Let F and G be formulas. Let σ be a good substitution for F, and let τ be a good substitution for G. Assume that $\sigma F \equiv \tau G$. Then F and G have the same principal logical symbol, and if F' is the k-th immediate subformula of F, and G' is the k-th immediate subformula of G, then $\sigma F' \equiv \tau G'$.

If A and B are two atomic formulas then let $\mathrm{equ}[A, B]$ denote the set consisting of the two clauses $\neg A \vee B$ and $A \vee \neg B$. So the set of clauses $\mathrm{equ}[A, B]$ asserts that A and B are equivalent.

Lemma 3.4.4
Let F and G be two formulas and let σ be a good substitution for F and τ a good substitution for G such that $\sigma F \equiv \tau G$. Let F_i $(i = 1, \ldots, k)$ be the immediate subformulas[7] of F, and let G_i $(i = 1, \ldots, k)$ be the immediate subformulas of G. Then each of the two clauses of $\mathrm{equ}[\sigma L_F, \tau L_G]$, *is derivable from*

$$\mathrm{def}_F \cup \mathrm{def}_G \cup \mathrm{equ}[\sigma L_{F_1}, \tau L_{G_1}] \cup \ldots \cup \mathrm{equ}[\sigma L_{F_k}, \tau L_{G_k}]$$

in resolution with subsumption. The derivation has a C-complexity of at most 4, *an L-complexity of at most* 9, *and a size of at most* $8 \cdot |\sigma F| + 12$. *Its clauses have cardinalities less than or equal to* 3.

[7]Here $k = 1$ if the principal logical symbol of F is $\neg$, $\exists$ or $\forall$; and $k = 2$ if the principal logical symbol of F is $\vee$ or $\wedge$.

Proof: First case: Assume $F \equiv F_1 \wedge F_2$ and $G \equiv G_1 \wedge G_2$. Then the following propositions hold.

$$\begin{array}{rcl} \neg L_F \vee L_{F_1}, \neg L_F \vee L_{F_2} & \in & \mathrm{def}_F \\ \neg L_{G_1} \vee \neg L_{G_2} \vee L_G & \in & \mathrm{def}_G \\ \neg\sigma L_{F_1} \vee \tau L_{G_1} & \in & \mathrm{equ}[\sigma L_{F_1}, \tau L_{G_1}] \\ \neg\sigma L_{F_2} \vee \tau L_{G_2} & \in & \mathrm{equ}[\sigma L_{F_2}, \tau L_{G_2}] \end{array}$$

Now, let us define

$$\begin{array}{rcl} d_1 & \stackrel{\mathrm{def}}{=} & \neg\sigma L_F \vee \tau L_{G_1} \\ d_2 & \stackrel{\mathrm{def}}{=} & \neg\sigma L_F \vee \tau L_{G_2} \\ d_3 & \stackrel{\mathrm{def}}{=} & \neg\sigma L_F \vee \neg\tau L_{G_2} \vee \tau L_G \\ d_4 & \stackrel{\mathrm{def}}{=} & \neg\sigma L_F \vee \tau L_G. \end{array}$$

Then d_1 is subsumed by a resolvent of $\neg L_F \vee L_{F_1}$ and $\neg\sigma L_{F_1} \vee \tau L_{G_1}$. Similarly, d_2 is subsumed by a resolvent of $\neg L_F \vee L_{F_2}$ and $\neg\sigma L_{F_2} \vee \tau L_{G_2}$. The clause d_3 is subsumed by a resolvent of $\neg L_{G_1} \vee \neg L_{G_2} \vee L_G$ and d_1. Finally, the clause d_4 is subsumed by a resolvent of d_2 and d_3. The size of the derivation $d_1\#d_2\#d_3\#d_4\#$ is equal to $4 \cdot |\sigma L_F| + 2 \cdot |\tau L_G| + |\tau L_{G_1}| + 2 \cdot |\tau L_{G_2}| + 5 \cdot |\neg| + 5 \cdot |\vee| + 4 \cdot |\#|$. From Lemma 3.4.1 it follows that $|\sigma L_F| \le |\sigma F|$, that $|\tau L_G| \le |\tau G|$, and that $|\tau L_{G_1}| + |\tau L_{G_2}| + 1 \le |\tau G_1| + |\tau G_2| + 1 = |\tau G|$. So the size of the derivation of d_4 is at most $4 \cdot |\sigma F| + 4 \cdot |\tau G| + 12$ which is equal to $8 \cdot |\sigma F| + 12$ since $\sigma F \equiv \tau G$. The C-complexity is 4, and the L-complexity is 9. The proof for the derivation of the clause $\sigma L_F \vee \neg\tau L_G$ is quite analogous.

Second case: Assume $F \equiv \exists x I$ and $G \equiv \exists x J$. Let $x_1, \ldots, x_n$ be the variables occurring free in F. Let ρ be the substitution $\{x \leftarrow f_F(x_1, \ldots, x_n)\}$ where f_F is the Skolem function associated with F (see the definition in Section 2.2). Then the following propositions hold.

$$\begin{array}{rcl} \neg L_F \vee \rho L_I & \in & \mathrm{def}_F \\ \neg L_J \vee L_G & \in & \mathrm{def}_G \\ \neg\sigma L_I \vee \tau L_J & \in & \mathrm{equ}[\sigma L_I, \tau L_J]. \end{array}$$

Now let

$$\begin{array}{rcl} d_1 & \stackrel{\mathrm{def}}{=} & \neg\sigma L_I \vee \tau L_G \\ d_2 & \stackrel{\mathrm{def}}{=} & \neg\sigma L_F \vee \tau L_G. \end{array}$$

Then d_1 is subsumed by a resolvent of $\neg L_J \vee L_G$ and $\neg\sigma L_I \vee \tau L_J$. Now, let υ be the substitution $\{x \leftarrow f_F(\sigma x_1, \ldots, \sigma x_n)\}$. We have assumed that σ is a good substitution for F. Moreover, x occurs bound in F. It follows that $\sigma\rho = \upsilon\sigma$ which can be seen by an easy calculation[8]. So $\sigma\rho L_I = \upsilon\sigma L_I$. Thus the clause

[8]If y is a variable that is in the domain of σ then the variable x does not occur in the term σy since σ is good for F. If, on the other hand, y is not in the domain of σ then $\sigma y = y$. If $y \neq x$

$d \stackrel{\text{def}}{=} \neg\sigma L_F \vee \upsilon\tau L_G$ is subsumed by a resolvent of the clauses $\neg L_F \vee \rho L_I$ and d_1. But the bound variable x does not occur in L_G. Since τ is a good substitution for G, the variable x therefore does not occur in τL_G either. So the clause d is identical to the clause d_2. So $d_1\#d_2\#$ is a derivation of the clause $\neg\sigma L_F \vee \tau L_G$ from the clause set $\text{def}_F \cup \text{def}_G \cup \text{equ}[\sigma L_I, \tau L_J]$ by resolution with subsumption. The C-complexity of the derivation is 2, and its L-complexity is 4. The size of the derivation is at most $2 \cdot |\sigma F| + 2 \cdot |\tau G| + 2$, i.e., $4 \cdot |\sigma F| + 2$. The estimate goes as in the first case. Note that no Skolem function enters into any resolvents. The estimate for the derivation of $\sigma L_F \vee \neg\tau L_G$ runs the same way.

The proofs for all the other cases follow the same pattern.

q.e.d.

Proposition 3.4.1
Let F and G be two formulas. Let σ be a good substitution for F and let τ be a good substitution for G, such that $\sigma F \equiv \tau G$. Let n be the structural complexity of F. Then there is a derivation $\mathcal{Q}$ of the clause set $\text{equ}[\sigma L_F, \tau L_G]$ from the clause set $\text{def}_F \cup \text{def}_G$ by resolution with subsumption such that the following statements hold.

$$\begin{aligned} |\mathcal{Q}|_C &\leq 8n \\ |\mathcal{Q}|_L &\leq 18n \\ |\mathcal{Q}| &\leq (16 \cdot |\sigma F| + 24) \cdot n. \end{aligned}$$

The cardinalities of the clauses of $\mathcal{Q}$ are less than or equal to 3.

Proof: The proof is by induction on the structural complexity of F.

If F is an atomic formula then $\neg\sigma L_F \vee \tau L_G$ is subsumed by a resolvent of the clause $\neg L_F \vee F \in \text{def}_F$ with the clause $L_G \vee \neg G \in \text{def}_G$. The size of the derivation is $|\neg\sigma L_F \vee \tau L_G| + 1$ which is less than or equal to $2 \cdot |\sigma F| + 3$. For the other clause $\sigma L_F \vee \neg\tau L_G$ there is a derivation of the same length. So the size of the derivation of the clause set $\text{equ}[\sigma L_F, \tau L_G]$ from the clause set $\text{def}_F \cup \text{def}_G$ by resolution with subsumption is at most $4 \cdot |\sigma F| + 6$. Its C-complexity is 2, and its L-complexity is 4. The structural complexity n of F is 1.

For the induction step, Lemma 3.4.4 implies that there is a derivation $\mathcal{Q}'$ of $\text{equ}[\sigma L_F, \tau L_G]$ from the clause set

$$\text{def}_F \cup \text{def}_G \cup \text{equ}[\sigma L_{F_1}, \tau L_{G_1}] \cup \ldots \cup \text{equ}[\sigma L_{F_k}, \tau L_{G_k}]$$

in resolution with subsumption whose C-complexity is at most 8, whose L-complexity is at most 18, and whose size is at most $16 \cdot |\sigma F| + 24$. For $j = 1, \ldots, k$,

then, in both cases, the variable x does not occur in σy. Therefore, if $y \neq x$ then $\sigma y = \upsilon\sigma y$. But, if $y \neq x$ then also $\rho y = y$ and therefore $\sigma\rho y = \sigma y$. So, if $y \neq x$ then $\sigma\rho y = \upsilon\sigma y$. On the other hand, x is not in the domain of σ since x is bound in F, and since σ is good for F. Therefore $\sigma x = x$. But, by definition, $\sigma\rho x = \upsilon x$. So $\sigma\rho x = \upsilon\sigma x$. So, in summary we have $\sigma\rho y = \upsilon\sigma y$ for all variables y.

let n_j be the structural complexity of the j-th immediate subformula F_j of F. Then $n = n_1 + \cdots + n_k + 1$. By induction hypothesis, for each $j = 1, \ldots, k$, there exists a derivation $\mathcal{Q}_j$ of $\text{equ}[\sigma L_{F_j}, \tau L_{G_j}]$ from $\text{def}_F \cup \text{def}_G$ whose C-complexity is at most $8n_j$, whose L-complexity is at most $18n_j$, and whose size is at most $(16 \cdot |\sigma F_j| + 24)n_j$. Now let $\mathcal{Q} \stackrel{\text{def}}{=} \mathcal{Q}_1 \ldots \mathcal{Q}_k \mathcal{Q}'$. Then an easy calculation shows that the required inequalities hold.

q.e.d.

3.5 Transformation of Sequents to Clauses

Let $\mathcal{D}$ be an arbitrary derivation in the sequent calculus. Then we give in this section a transformation of the sequents of $\mathcal{D}$ to clauses. If $\mathcal{D}$ is a derivation of a formula F in the sequent calculus, then the transformation will be such that the transform of the endsequent $\Longrightarrow F$ of $\mathcal{D}$ is L_F. In Section 3.6, we shall establish the p-simulatability of the sequent calculus in extended definitional resolution. This means that we shall prove that there is a short derivation of L_F from def_F in extended resolution. The idea of the proof is the following.

First we prove that there is a short ER-derivation of the transform of any axiom of $\mathcal{D}$ from def_F. If c is a transform of the conclusion of a rule application R, then we prove that there is a short ER-derivation of c from transforms of premises of R and from def_F. Now, the transformation will be such that, if $\mathcal{D}$ is a tree derivation, then each occurrence of a sequent in $\mathcal{D}$ has one and only one transform. In this case we can combine the obtained ER-derivations into a short ER-derivation of L_F from def_F, and we are done. If $\mathcal{D}$ is not a tree derivation, however, there may not be a uniquely defined transform for a given sequent occurrence. Rather, the number of possible transforms for a sequent occurrence χ may grow exponentially with respect to the size of the derivation. Our remedy for this problem will be not to derive all transforms of χ, but to choose one transform of χ arbitrarily, and to prove that there is a short derivation of any transform of χ from any other transform of χ.

3.5.1 The Definition Set for a Derivation in the Sequent Calculus

In Section 3.6, we shall prove that, for every derivation $\mathcal{D}$ of a formula F in the sequent calculus, there is a derivation $\mathcal{X}$ of L_F from def_F in extended resolution, and that the size of $\mathcal{X}$ is bounded by a polynomial of the size of $\mathcal{D}$. The derivation $\mathcal{X}$ will consist of two parts. In the first part there will only be extension steps, and in the second part there will only be resolution and factorization steps. The set of clauses obtained by the extension steps will be denoted by $\text{def}_{\mathcal{D}}$, and it is defined as follows.

Definition 3.5.1
If $\mathcal{D}$ is a derivation of a sequent $\mathcal{S}$ in the sequent calculus, then the set of clauses $\text{def}_\mathcal{D}$ is defined to be the union of all sets def_F, such that F is a member of the antecedent or of the succedent of $\mathcal{S}$, or F is the cut formula of some cut in $\mathcal{D}$. We call def_F the *definition set* of $\mathcal{D}$.

Definition 3.5.2
By the *cut-complexity* of a derivation $\mathcal{D}$ in the sequent calculus we mean the integer $\sum_{\kappa\in K} l_\kappa$ where K is the set of all occurences of a cut in $\mathcal{D}$, and, for each $\kappa \in K$, the integer l_κ is the structural complexity of the cut formula of κ.

Proposition 3.5.1
Let F be a formula and $\mathcal{D}$ a derivation of F in the sequent calculus. Let n be the cut complexity of $\mathcal{D}$. Let N be the maximum of the sizes of all cut formulas in $\mathcal{D}$. Then there is a derivation $\mathcal{E}$ of $\text{def}_\mathcal{D}$ *from* def_F *by at most n extensions. Moreover, the following inequalities hold.*

$$\begin{aligned} |\mathcal{E}|_C &\leq 3n \\ |\mathcal{E}|_L &\leq 7n \\ |\mathcal{E}| &\leq (5N+8)n \end{aligned}$$

Proof: This follows from Proposition 3.2.2 by induction on the number of applications of the cut rule in $\mathcal{D}$.

q.e.d.

So it suffices to prove that a short resolution derivation of L_F from $\text{def}_\mathcal{D}$ exists. We shall define a transform $(c_1, \ldots, c_r)$ of $\mathcal{D}$. This transform will be a finite sequence of clauses, and we shall prove that a short derivation of c_j from $\text{def}_\mathcal{D} \cup \{c_1, \ldots, c_{j-1}\}$ by resolution with subsumption exists for all $j = 1, \ldots, r$. The last clause c_r of the transform of $\mathcal{D}$ will be the clause $\{L_F\}$. Then we shall concatenate all these derivations to obtain a derivation of L_F from $\text{def}_\mathcal{D}$ by resolution with subsumption. Finally, we shall show that this derivation can be transformed to a short resolution derivation of L_F from $\text{def}_\mathcal{D}$, and therefore also to a short derivation of L_F from def_F in extended resolution.

3.5.2 Transforms

In this section we introduce the concepts of a transform of an occurrence of a formula, of a transform of an occurrence of a sequent, and of a transform of a derivation.

Definition 3.5.3
Let R be a rule application in a derivation $\mathcal{D}$ in the sequent calculus. Let η be an occurrence as a subformula in one of the premises of R, such that η has an immediate subordinate in the conclusion of R. Then the *characteristic substitution* σ for η in R is defined as follows.

1. If η is a suboccurrence of an $F\{x\backslash a\}$-occurrence, then $\sigma \stackrel{\text{def}}{=} \{x \leftarrow a\}$.
2. If η is a suboccurrence of an $F\{x\backslash t\}$-occurrence, then $\sigma \stackrel{\text{def}}{=} \{x \leftarrow t\}$.
3. Otherwise, σ is the empty substitution.

Lemma 3.5.1
Let η be an occurrence of a formula as a subformula in the premise of a rule application R in a derivation. Let θ be the immediate subordinate of η in R, and let σ be the characteristic substitution for η in R. Then $\tilde{\eta} \equiv \sigma\tilde{\theta}$.

Lemma 3.5.2
If a formula occurrence θ is a subordinate of a formula occurrence η in a derivation in the sequent calculus then there is a good substitution σ for $\tilde{\theta}$ such that $\tilde{\eta} \equiv \sigma\tilde{\theta}$.

Proof: Under the hypothesis of the lemma, there is a sequence $(R_1, \ldots, R_r)$ of rule applications, and a sequence $\eta_0, \ldots, \eta_r$ of occurrences of formulas, such that $\eta_0 = \eta$, and η_{s-1} is a suboccurrence of a premise of R_s, and η_s is a suboccurrence of the conclusion of R_s for $s = 1, \ldots, r$, and $\eta_r = \theta$. Now, let σ_s be the characteristic substitution for η_{s-1} in R_s for all $s = 1, \ldots, r$. Let $\sigma \stackrel{\text{def}}{=} \sigma_1 \ldots \sigma_r$. It follows from the last lemma by induction that $\tilde{\eta} \equiv \sigma\tilde{\theta}$. Moreover, an induction argument yields that the set of variables occurring bound in η_s equals the set of variables occurring bound in θ. So, no variables occurring bound in θ are replaced by any of the substitutions $\sigma_1, \ldots, \sigma_r$. It follows that σ is a good substitution for $\tilde{\theta}$.

q.e.d.

Definition 3.5.4
Let $p = (\eta_0, \ldots, \eta_n)$ be a path for an occurrence η_0 of a formula in a derivation $\mathcal{D}$. For $j = 1, \ldots, n$, let σ_j be the characteristic substitution for η_{j-1} in the j-th rule application of the path p. Then the *characteristic substitution* for p is defined to be the substitution $\sigma_1 \ldots \sigma_n$.

Definition 3.5.5
If $p = (\eta_0, \ldots, \eta_n)$ is a path in a derivation then the formula $\tilde{\eta}_n$ is called the *characteristic formula* for p. If p is a path in a derivation and F is the characteristic formula for p then L_F is called the *characteristic literal* for p.

Definition 3.5.6
Let $\mathcal{D}$ be a derivation in the sequent calculus. Let α be an occurrence of a formula as a member of a sequent of $\mathcal{D}$. Let p be a path for α in $\mathcal{D}$. Let L be the characteristic literal for p, and let σ be the characteristic substitution for p. Then the *transform* of α for p (with respect to $\mathcal{D}$) is the literal σL.

Lemma 3.5.3
Let $\mathcal{D}$ be a derivation in the sequent calculus. Let α be an occurrence of a formula as a member of a sequent of $\mathcal{D}$. Then α has at least one transform. If $\mathcal{D}$ is a tree derivation then α has one and only one transform.

Proof: By Lemma 3.3.1, there is a path for α. So there is a transform of α for this path. If $\mathcal{D}$ is a tree derivation, then, by Lemma 3.3.4, there is only one path for α. So there is only one transform of α.

q.e.d.

If k is the number of subformulas of members of the endsequent of the derivation plus the number of subformulas of cut formulas in the derivation, then the number of transforms of any given occurrence α of a formula is less than or equal to k, which is in turn less than the size of the derivation. Another upper bound for the number of transforms of α is the number of paths for α. This bound is equal to 1 in a tree derivation, but it is exponential in general.

Lemma 3.5.4

Let $\mathcal{D}$ be a derivation in the sequent calculus. Let α be an occurrence of a formula as a member of a sequent of $\mathcal{D}$. Let K be a transform of α. Then the following statements hold.

1. *There is exactly one formula F such that there exists a path p for which K is a transform of α and for which F is the characteristic formula.*

2. *There is exactly one literal L such that there exists a path p for which K is a transform of α and for which L is the characteristic literal.*

3. *There is exactly one substitution σ such that there exists a path p for which K is a transform of α and for which σ is the characteristic substitution.*

Proof: Let F be the characteristic formula, L the characteristic literal, and σ the characteristic substitution for a path p for which K is a transform of α. Then $K \equiv \sigma L_F$. So, the predicate symbol of the literal K is P_F. So, P_F and therefore also F is uniquely determined by K. So the characteristic literal $L \equiv L_F$ is also uniquely determined by K. Since the domain of σ is the set of variables in L, and since $K \equiv \sigma L$, the substitution σ is also uniquely determined by K.

q.e.d.

Definition 3.5.7

Let $\mathcal{D}$ be a derivation in the sequent calculus. Let χ be an occurrence of a sequent in $\mathcal{D}$. Let $\alpha_1, \ldots, \alpha_m$ be all the occurrences of formulas as members in the antecedent of χ, and let $\beta_1, \ldots, \beta_n$ be all the occurrences of formulas as members in the succedent of χ. Let $K_1, \ldots, K_m$ be transforms of $\alpha_1, \ldots, \alpha_m$, respectively, and let $L_1, \ldots, L_n$ be transforms of $\beta_1, \ldots, \beta_n$. Then the clause $\neg K_1 \vee \ldots \vee \neg K_m \vee L_1 \vee \cdots \vee L_n$ is said to be a *transform* of χ.

The following lemma is an immediate consequence of Lemma 3.5.3.

Lemma 3.5.5
Let $\mathcal{D}$ be a derivation in the sequent calculus. Let χ be an occurrence of a sequent in $\mathcal{D}$. Then χ has at least one transform. If $\mathcal{D}$ is a tree derivation then χ has one and only one transform.

If k is the number of subformulas of members of the endsequent of the derivation plus the number of subformulas of cut formulas in the derivation, then the number of transforms of any given occurrence χ of a sequent is at most k^n, where n is the cardinality of the sequent $\tilde{\chi}$. Also, the number of transforms of χ is at most p^n where p is the number of ways from the occurrence χ to the occurrence of the endsequent in the derivation dag. The first of these bounds is exponential in the size of the derivation, the second one is possibly even worse. But, in the case of a tree derivation, the second bound equals 1.

Definition 3.5.8
Let $\mathcal{D}$ be a derivation in the sequent calculus, and let $(\chi_1, \ldots, \chi_r)$ be the sequence of the occurrences of its sequents. Let $c_1, \ldots, c_r$ be transforms of $\chi_1, \ldots, \chi_r$, respectively. Then the sequence of clauses $(c_1, \ldots, c_n)$ is called a *transform* of $\mathcal{D}$.

Lemma 3.5.6
Every derivation in the sequent calculus has a transform. Every tree derivation in the sequent calculus has one and only one transform.

Lemma 3.5.7
If $\mathcal{D}$ is a derivation of a formula F in the sequent calculus then the last clause of any transform of $\mathcal{D}$ is the unit clause L_F.

Lemma 3.5.8
If F is the characteristic formula of a path in a derivation $\mathcal{D}$ in the sequent calculus, then $\mathrm{def}_F \subseteq \mathrm{def}_{\mathcal{D}}$.

Lemma 3.5.9
Let M be the size of a sequent $\mathcal{S}$, and let $N_1, \ldots, N_k$ be the respective sizes of the members of $\mathcal{S}$. Then $\sum_{i=1}^{k} N_i \leq M - k + 1$.

Lemma 3.5.10
Let $\mathcal{D}$ be a derivation and let χ be an occurrence of a sequent $\mathcal{S}$ in $\mathcal{D}$. Let k be the number of immediate suboccurrences of χ, and let M be the size of $\mathcal{S}$. Let w be the word representation of a transform of χ. Then the size of the word $w\#$ is at most equal to $M + k + 1$.

Proof: Let $N_1, \ldots, N_k$ be the respective sizes of the members of $\mathcal{S}$. Let $n_1, \ldots, n_k$ be the respective sizes of their transforms. From Lemma 3.4.1 it follows that $n_i \leq N_i$ for $i = 1, \ldots, k$. So, Lemma 3.5.9 yields $\sum_{i=1}^{k} n_i \leq M - k + 1$. But the size of $w\#$ equals $\sum_{i=1}^{k} n_i$ plus the number of occurrences of $\neg$, $\vee$ and $\#$ in $w\#$. Unless $\mathcal{S}$ is empty, there are exactly $k - 1$ occurrences of $\vee$, exactly 1 occurrence of $\#$, and at most k occurrences of $\neg$. These are altogether at most $2k$ occurrences of these three symbols. So the total size of $w\#$ is at most $(M - k + 1) + 2k$.

q.e.d.

3.5.3 Equivalence of Transforms

Let λ be either an occurrence of a formula as a member of a sequent in a derivation, or an occurrence of a sequent in a derivation. Let X and Y be two transforms of λ. We shall prove in this section that then there is a short derivation of Y from $\mathrm{def}_{\mathcal{D}} \cup \{X\}$ in resolution with subsumption.

Proposition 3.5.2
Let α and β be occurrences of a formula F as members of sequents of a derivation in the sequent calculus. Let A be a transform of α, and let B be a transform of β. Let n be the structural complexity of F, and let N be the size of F. Then there is a derivation $\mathcal{Q}$ of the clause set $\mathrm{equ}[A, B]$ from the clause set $\mathrm{def}_{\mathcal{D}}$ by resolution with subsumption such that the following statements hold.

$$\begin{aligned} |\mathcal{Q}|_C &\leq 8n \\ |\mathcal{Q}|_L &\leq 18n \\ |\mathcal{Q}| &\leq (16N + 24)n. \end{aligned}$$

The cardinalities of the clauses of $\mathcal{Q}$ are less than or equal to 3.

Proof: Since A is a transform of α, and B is a transform of β, there is a path p for α such that A is the transform of α for p, and there is a path q for β such that B is the transform of β for q. Let G be the characteristic formula for p, and let σ be the characteristic substitution for p. Let H and τ be the characteristic formula and substitution, respectively, for q. Then $\mathrm{def}_G \cup \mathrm{def}_H \subseteq \mathrm{def}_{\mathcal{D}}$ by Lemma 3.5.8. Moreover, $F \equiv \sigma G$ and $F \equiv \tau H$, and σ is good for G, and τ is good for H. So the structural complexity of G equals the structural complexity of F which is n. Furthermore, we have $A \equiv \sigma L_G$ and $B \equiv \tau L_H$. The assertion follows now from Proposition 3.4.1.

q.e.d.

We shall exploit this result in two ways as follows.

Proposition 3.5.3
Let $\mathcal{D}$ be a derivation in the sequent calculus, and let χ be an occurrence of a sequent $H \Longrightarrow H$ as an axiom of $\mathcal{D}$. Let c be a transform of χ. Let n be the structural complexity of H. Let N be the size of H. Then there is a derivation $\mathcal{Q}$ of c from $\mathrm{def}_{\mathcal{D}}$ by resolution with subsumption such that the following inequalities hold.

$$\begin{aligned} |\mathcal{Q}|_C &\leq 8n \\ |\mathcal{Q}|_L &\leq 18n \\ |\mathcal{Q}| &\leq (16N + 24)n. \end{aligned}$$

The cardinalities of the clauses of $\mathcal{Q}$ are less than or equal to 3.

Proof: Let α and β be the two occurrences of H as members of χ. Then Proposition 3.5.2 yields the assertion.

q.e.d.

Proposition 3.5.4

Let χ be an occurrence of a sequent $\mathcal{S}$ in a derivation in the sequent calculus. Let the clauses c and d be two transforms of χ. Let k be the number of members of $\mathcal{S}$, and let M be the size of $\mathcal{S}$. Then there is a derivation $\mathcal{Q}$ of the clause d from the clause set $\mathrm{def}_{\mathcal{D}} \cup \{c\}$ *by resolution with subsumption such that the following inequalities hold.*

$$\begin{aligned} |\mathcal{Q}|_C &\leq 8M \\ |\mathcal{Q}|_L &\leq 18M + k^2 \\ |\mathcal{Q}| &\leq 32M^2 \end{aligned}$$

The cardinalities of the clauses of $\mathcal{Q}$ are less than or equal to $\max(3, k)$.

Proof: Let $\alpha_1, \ldots, \alpha_k$ be the immediate suboccurrences of χ. Let $F_i \stackrel{\text{def}}{=} \tilde{\alpha_i}$ for all $i = 1, \ldots, k$. Let A_i be the transform of α_i contained in c, and let B_i be the transform of α_i contained in d, for each $i = 1, \ldots, k$. So $c = \{\neg A_1, \ldots, \neg A_l, A_{l+1}, \ldots, A_k\}$, and $d = \{\neg B_1, \ldots, \neg B_l, B_{l+1}, \ldots, B_k\}$ for some $l \leq k$. Let us denote these literals by $K_1, \ldots, K_k$ and by $L_1, \ldots, L_k$, respectively. Let n be the maximal structural complexity of members of $\mathcal{S}$. Let N_i be the size of F_i, and let n_i be the structural complexity of F_i, for $i = 1, \ldots, k$. By Proposition 3.5.2, there is a derivation $\mathcal{Q}_i$ of the clause set $\mathrm{equ}[A_i, B_i]$ from the clause set $\mathrm{def}_{\mathcal{D}}$ by resolution with subsumption such that the following statements hold.

$$\begin{aligned} |\mathcal{Q}_i|_C &\leq 8n_i \\ |\mathcal{Q}_i|_L &\leq 18n_i \\ |\mathcal{Q}_i| &\leq (16N_i + 24)n_i. \end{aligned}$$

Moreover, from these clause sets and the clause $c = \{K_1, \ldots, K_n\}$, the clauses $\{L_1, \ldots, L_i, K_{i+1}, \ldots, K_k\}$ can be derived successively for $i = 1, \ldots, k$ until finally d is derived. Each such step amounts to one step by resolution with subsumption. Let us denote this derivation by $\mathcal{Q}'$. Since each of the k clauses in this derivation is a transform of χ, it follows from Lemma 3.5.10 that

$$\begin{aligned} |\mathcal{Q}'|_C &= k \\ |\mathcal{Q}'|_L &\leq k^2 \\ |\mathcal{Q}'| &\leq (M + k + 1)k. \end{aligned}$$

The following inequalities hold[9]. $n_i \leq N_i - 2$. By Lemma 3.5.9, we have $\sum_{i=1}^{k} N_i \leq M - k + 1$. Therefore $\sum_{i=1}^{k} n_i \leq M - 3k + 1$. It follows that $\sum_{i=1}^{k} n_i \leq M - 2k$.

[9] Here we assume that the formulas are fully parenthesized. So, if P is a nullary predicate symbol then we assume that the atomic formula $P()$ is not abbreviated as P.

Now, let $\mathcal{Q}$ be the derivation $\mathcal{Q}_1 \ldots \mathcal{Q}_k \mathcal{Q}'$. Then

$$|\mathcal{Q}|_C \leq 8 \sum_{i=1}^{k} n_i + k \leq 8M$$
$$|\mathcal{Q}|_L \leq 18 \sum_{i=1}^{1} n_i + k^2 \leq 18M + k^2$$
$$|\mathcal{Q}| \leq \sum_{i=1}^{k} (16N_i + 24)n_i + (M + k + 1)k.$$

It follows therefore that $|\mathcal{Q}| \leq (16(M - k + 1) + 24k)n + (M + k + 1)k$. But this is less than or equal to $(M + k + 1)(16n + k)$. Now, $k + 1 \leq M$, and $n + k \leq M$. So $|\mathcal{Q}| \leq 32M^2$.

q.e.d.

3.5.4 Inheritance of Transforms

We have seen that, for tree derivations, there is a unique transform of any occurrence of a literal or of a sequent. For general derivations, this is not true. There is, however, a natural way to propagate the choice, which transform to take, from the endsequent along a path back to the axioms. We shall call this *inheritance*. We shall see that, if the transforms of premises of a rule application are inherited from the transform of its conclusion, then there is a short resolution refutation of the transform of the conclusion from the transforms of the premises and from the clause set $\text{def}_{\mathcal{D}}$. The inheritance relations along different paths are, in general, not compatible with each other. But we shall use Proposition 3.5.4 to show nevertheless that any proof in the sequent calculus can be transformed at polynomial cost to a derivation in extended definitional resolution.

Definition 3.5.9
Let $\mathcal{D}$ be a derivation in the sequent calculus. Let $p = (\eta_0, \ldots, \eta_n)$ and $q = (\theta_1, \ldots, \theta_n)$ be paths in $\mathcal{D}$ such that η_j is a suboccurrence of θ_j for $j = 1, \ldots, n$. Then we say that p is an *extension* of q.

Definition 3.5.10
Let R be an application of a rule in a derivation $\mathcal{D}$ in the sequent calculus. Let α be an occurrence of a formula as a member of one of the premises of R, and let β be the immediate descendant of α in R. Let L be a transform of β for a path q, and let K be a transform of α for an extension p of q. Then we say that the transform K of α is *inherited* from the transform L of β via R.

Lemma 3.5.11
Let R be an application of a rule in a derivation $\mathcal{D}$ in the sequent calculus. Let α be an occurrence of a formula as a member of one of the premises of R, and let β be the immediate descendant of α in R. Let L be a transform of β. Then there is one and only one transform K of α which is inherited from the transform L of β via R. Moreover, there is a substitution σ, and there are formulas F and G such that the following statements hold.

1. *If β is not the principal occurrence of R then $K \equiv L \equiv \sigma L_F$.*

2. *If R is an application of $\neg$-IA or of $\neg$-IS, and if β is the principal occurrence of R, then $K \equiv \sigma L_F$ and $L \equiv \sigma L_{\neg F}$.*

3. *If R is an application of $\circ$-IA or of $\circ$-IS where $\circ$ is a binary propositional connective, and if β is the principal occurrence of R, then $K \equiv \sigma L_F$ or $K \equiv \sigma L_G$, and $L \equiv \sigma L_{F \circ G}$.*

4. *If R is an application of $\forall$-IA, $\forall$-IS, $\exists$-IA, or $\exists$-IS then $K \equiv \rho\sigma L_F$, and $L \equiv \sigma L_{\forall x F}$ or $L \equiv \sigma L_{\exists x F}$ (depending upon whether R is an application of a $\forall$-rule or of an $\exists$-rule), where the substitution ρ is the characteristic substitution of R (i.e., $\rho \stackrel{\text{def}}{=} \{x \leftarrow t\}$ in the cases of $\forall$-IA and $\exists$-IS, and $\rho \stackrel{\text{def}}{=} \{x \leftarrow a\}$ in the cases of $\forall$-IS and $\exists$-IA).*

Proof: By Lemma 3.5.4, the transform L of β uniquely determines its characteristic formula, say H, its characteristic literal L_H, and its characteristic substitution, say σ. Now, $L \equiv \sigma L_H$. The immediate subordinate, say η, of α is a suboccurrence of β. Since $\tilde{\beta} \equiv \sigma H$, there is a subformula E of H, exactly corresponding to the suboccurrence η of β, such that $\tilde{\alpha} \equiv \rho\sigma E$. The characteristic formula E and the characteristic substitution $\rho\sigma$ for a path p for α are independant of p as long as p is an extension of a path for β for which L is a transform of β. So, also the transform $K \equiv \rho\sigma L_E$ of α inherited from the transform L of β via R, is independant of the path p. Now, $E \equiv H$ if β is not a principal occurrence of R, and E is an immediate subformula of H if β is a principal occurrence of R. The assertion follows now directly in all cases.

q.e.d.

Definition 3.5.11
Let R be a rule application in a derivation, and let χ be a premise of R, and let ψ be the conclusion of R. We say that a transform c of χ is *inherited* from a transform d of ψ if the transform of each immediate suboccurrence of χ having an immediate descendant in R, is inherited from the transform of its immediate descendant in R.

Lemma 3.5.12
Let R be a rule application, let $\chi_1, \ldots, \chi_k$ be the premises of R, and let ψ be the conclusion of R. Let d be a transform of ψ, and let $c_1, \ldots, c_k$ be transforms of $\chi_1, \ldots, \chi_k$, respectively, which are inherited from d via R. Then the respective transforms of all A-occurrences in R are identical to each other. The same holds for B-, D-, and E-, Δ-, Γ-, Θ-, and Λ-occurrences.

Proof: This can be read off directly from the last lemma.

q.e.d.

Notation for clauses
If $S_1, \ldots, S_j$ and $T_1, \ldots, T_k$ are finite sets of atomic formulas then we denote by

$$S_1, \ldots, S_j \Longrightarrow^* T_1, \ldots, T_k$$

the clause

$$\{\neg L \mid L \in S_1 \cup \cdots \cup S_j\} \cup T_1 \cup \cdots \cup T_k.$$

In this notation we also allow to write atomic formulas instead of sets of atomic formulas. An atomic formula L is understood to stand for the singleton set $\{L\}$. So, for example, $\{J, K\} \Longrightarrow^* \{L\}, M$ denotes the clause $\{\neg J, \neg K, L, M\}$.

We shall now consider a given rule application R in a derivation in the sequent calculus. Furthermore, we assume that we are given a transform d of the conclusion of R, and the transforms inherited from d via R for the premises of R. We consider this rule application and these transforms as fixed for the following notation.

Notation
By A^* we denote the transform of the A-occurrences. Analogously, we define B^*, D^*, E^*, $(F\{x\backslash t\})^*$, $(F\{x\backslash a\})^*$, $(\forall x F)^*$, $(\exists x F)^*$, Γ^*, Δ^*, Θ^*, and Λ^*.

By the last lemma, this definition is sound. Now, the transforms of the premises or of the conclusion of our given rule application, can be denoted by taking the corresponding premise or conclusion of the formulation of the rule, and by appending an asterisk to each of its substrings denoting a member of a sequent, to each of its capital Greek letters, and to its sequent arrow. For example, if our rule is an $\wedge$-IS then the transforms of its two premises are the clauses $\Gamma^* \Longrightarrow^* \Theta^*, A^*$ and $\Gamma^* \Longrightarrow^* \Theta^*, B^*$, and the transform of its conclusion is the clause $\Gamma^* \Longrightarrow^* \Theta^*, (A \wedge B)^*$.

Lemma 3.5.13
Let $\mathcal{D}$ be a derivation in the sequent calculus, and let R be a rule application in $\mathcal{D}$. Assume we are given a transform d of the conclusion of R and the transforms of the premises of R inherited from d. Let us adopt the notation introduced above. Then each of the following clauses, provided it is defined, is subsumed by a clause in $\mathrm{def}_{\mathcal{D}}$.

For $\neg$-IA: $\qquad A^*, (\neg A)^* \Longrightarrow^*$

For $\neg$-IS: $\qquad \Longrightarrow^* A^*, (\neg A)^*$

For $\wedge$-IA: $\qquad (A \wedge B)^* \Longrightarrow^* A^* \qquad (A \wedge B)^* \Longrightarrow^* B^*$

For $\wedge$-IS: $\qquad A^*, B^* \Longrightarrow^* (A \wedge B)^*$

For $\vee$*-IA:* $$(A \vee B)^* \Longrightarrow^* A^*, B^*$$

For $\vee$*-IS:* $$A^* \Longrightarrow^* (A \vee B)^* \qquad B^* \Longrightarrow^* (A \vee B)^*$$

For $\rightarrow$*-IA:* $$(A \rightarrow B)^*, A^* \Longrightarrow^* B^*$$

For $\rightarrow$*-IS:* $$\Longrightarrow^* A^*, (A \rightarrow B)^* \qquad B^* \Longrightarrow^* (A \rightarrow B)^*$$

For $\forall$*-IA:* $$(\forall x F)^* \Longrightarrow^* (F\{x\backslash t\})^*$$

For $\forall$*-IS:* $$\{a \leftarrow T\}(F\{x\backslash a\})^* \Longrightarrow^* (\forall x F)^* \qquad \textit{for some term } T$$

For $\exists$*-IA:* $$(\exists x F)^* \Longrightarrow^* \{a \leftarrow T\}(F\{x\backslash a\})^* \qquad \textit{for some term } T$$

For $\exists$*-IS:* $$(F\{x\backslash t\})^* \Longrightarrow^* (\exists x F)^*$$

Proof: In the case of $\neg$-IA or $\neg$-IS, Lemma 3.5.11 implies that there is a substitution σ and a formula $\neg F$ at a terminal occurrence of $\mathcal{D}$, such that $A^* \equiv \sigma L_F$ and $(\neg A)^* \equiv \sigma L_{\neg F}$. So each of the first two clauses above is subsumed by a clause c in $\text{def}_{\neg F}$. But then $c \in \text{def}_{\mathcal{D}}$.

In the case of $\circ$-IA or $\circ$-IS where $\circ$ is a binary propositional connective, Lemma 3.5.11 implies that there is a substitution σ and a formula $F \circ G$ at a terminal occurrence of $\mathcal{D}$, such that $A^* \equiv \sigma L_F$ and $B^* \equiv \sigma L_G$ and $(A \circ B)^* \equiv \sigma L_{F \circ G}$. Again, each of the respective clauses in the list above is subsumed by a clause c in $\text{def}_{F \circ G}$. But then $c \in \text{def}_{\mathcal{D}}$.

In the case of a $\forall$-IA, Lemma 3.5.11 implies that there is a substitution σ and a formula $\forall x H$ at a terminal occurrence of $\mathcal{D}$, such that $(F\{x\backslash t\})^* \equiv \rho\sigma L_H$ where $\rho \stackrel{\text{def}}{=} \{x \leftarrow t\}$, and such that $(\forall x F)^* \equiv \sigma L_{\forall x H}$. But since x does not occur in $L_{\forall x H}$, it follows that $\rho\sigma L_{\forall x H} \equiv \sigma L_{\forall x H}$. Again, the respective clause in the list above is subsumed by a clause c in $\text{def}_{\forall x F}$. But then $c \in \text{def}_{\mathcal{D}}$. The case of an $\exists$-IS is quite analogous.

In the case of a $\forall$-IS, Lemma 3.5.11 implies that there is a substitution σ and a formula $\forall x H$ at a terminal occurrence of $\mathcal{D}$, such that $(F\{x\backslash a\})^* \equiv \rho\sigma L_H$ where $\rho \stackrel{\text{def}}{=} \{x \leftarrow a\}$, and such that $(\forall x F)^* \equiv \sigma L_{\forall x H}$. Let s be the term $f_{\forall x H}(x_1, \ldots, x_n)$ which occurs in the definition for the formula $\forall x F$ and which is built from its Skolem function. Furthermore, let T be the term σs. Then $\{x \leftarrow s\} L_H \Longrightarrow^* L_{\forall x H}$

is a clause in $\mathrm{def}_{\forall x H}$, and therefore in $\mathrm{def}_{\mathcal{D}}$. Now x does not occur in $L_{\forall x H}$ and is therefore not in the domain of σ. So an application of the substitution σ to this clause yields the clause $\{x \leftarrow T\}\sigma L_H \Longrightarrow^* \sigma L_{\forall x H}$. Since the eigenvariable a was assumed not to occur in $\forall x H$, it also does not occur in σL_H. So, $\{x \leftarrow T\}\sigma L_H \equiv \{a \leftarrow T\}\{x \leftarrow a\}\sigma L_H \equiv \{a \leftarrow T\}\rho\sigma L_H$. So this clause is identical to $\{x \leftarrow a\}(F\{x\backslash a\})^* \Longrightarrow^* (\forall x F)^*$. So we have proved that this clause is subsumed by a clause of $\mathrm{def}_{\mathcal{D}}$. The case of an $\exists$-IA is quite analogous.

q.e.d.

Proposition 3.5.5

Let $\mathcal{D}$ be a derivation in the sequent calculus, and let $R = (\chi_1, \ldots, \chi_r, \psi)$ be a rule application in $\mathcal{D}$. Let d be a transform of ψ, and let $c_1, \ldots, c_r$ be the transforms of $\chi_1, \ldots, \chi_r$ which are inherited from d via R. Let k be the cardinality of the conclusion, i.e., the number of occurrences of formulas as members of ψ, and let M be the size of the sequent $\tilde{\psi}$. Then there is a derivation $\mathcal{Q}$ of the clause d from the clause set $\mathrm{def}_{\mathcal{D}} \cup \{c_1, \ldots, c_r\}$ by resolution with subsumption, such that the following inequalities hold.

$$\begin{aligned} |\mathcal{Q}|_C &\leq 2 \\ |\mathcal{Q}|_L &\leq 2k+1 \\ |\mathcal{Q}| &< 3M. \end{aligned}$$

The cardinalities of the clauses of $\mathcal{Q}$ are less than or equal to $\max(3, k+1)$.

Proof: Let us denote the clause set $\mathrm{def}_{\mathcal{D}} \cup \{c_1, \ldots, c_r\}$ by S. We make a case distinction according to which rule of the sequent calculus we are dealing with.

1. In the case of the thinning rule, we have

$$c_1 = \Gamma^* \Longrightarrow^* \Theta^*$$

and

either $\quad d = D^*, \Gamma^* \Longrightarrow^* \Theta^* \quad$ or $\quad d = \Gamma^* \Longrightarrow^* \Theta^*, D^*$.

Then c_1 subsumes d. So $d\#$ is a derivation of d from S by resolution with subsumption. Its C-complexity is 1, its L-complexity is at most k, and, by Lemma 3.5.10, its size is at most $M + k + 1$, and therefore less than $3M$.

2. In the case of the contraction rule or the interchange rule, we have $c_1 = d$ (as sets of literals). So the empty word is a derivation of d from S. Its C-complexity, its L-complexity, and its size are 0.

3. In the case of the cut rule, we have

$$c_1 = \Gamma^* \Longrightarrow^* \Theta^*, D^* \qquad c_2 = D^*, \Delta^* \Longrightarrow^* \Lambda^*$$
$$d = \Gamma^*, \Delta^* \Longrightarrow^* \Theta^*, \Lambda^*.$$

The clause d is therefore subsumed by a resolvent of c_1 and c_2. So $d\#$ is a derivation of d from S. Its C-complexity is 1, its L-complexity is at most k, and, by Lemma 3.5.10, its size is at most $M + k + 1$.

4. In the case of $\neg$-IA, we have

$$c_1 = \Gamma^* \Longrightarrow^* \Theta^*, A^*$$

$$d = (\neg A)^*, \Gamma^* \Longrightarrow^* \Theta^*.$$

The clause d is therefore subsumed by a resolvent of c_1 and $(\neg A)^*, A^* \Longrightarrow^*$. But this latter clause is by Lemma 3.5.13 subsumed by a clause in $\mathrm{def}_{\mathcal{D}}$. So the clause d is subsumed by a resolvent of c_1 and of a clause in $\mathrm{def}_{\mathcal{D}}$. So $d\#$ is a derivation of d from S. Its C-complexity is 1, its L-complexity is at most k, and, by Lemma 3.5.10, its size is at most $M + k + 1$.

5. The cases of $\neg$-IS, $\wedge$-IA, and $\vee$-IS are quite analogous.

6. In the case of $\wedge$-IS, we have

$$c_1 = \Gamma^* \Longrightarrow^* \Theta^*, A^* \qquad c_2 = \Gamma^* \Longrightarrow^* \Theta^*, B^*$$

$$d = \Gamma^* \Longrightarrow^* \Theta^*, (A \wedge B)^*.$$

Now, let $e_0 \stackrel{\text{def}}{=} A^*, B^* \Longrightarrow^* (A \wedge B)^*$. Then, by Lemma 3.5.13, the clause e_0 is subsumed by some clause in $\mathrm{def}_{\mathcal{D}}$. Let $e \stackrel{\text{def}}{=} \Gamma^*, B^* \Longrightarrow^* \Theta^*, (A \wedge B)^*$. Then e is subsumed by a resolvent of c_1 and e_0, and d is subsumed by a resolvent of e and c_2. So $e\#d\#$ is a derivation of d from S by resolution with subsumption. Its C-complexity is 2, its L-complexity is at most $2k + 1$, and from Lemma 3.5.10 it follows that its size is at most $2(M + k + 1) + |B^*| + 1$. But the size of B^* is less than $M - 2k - 3$. So the size of the derivation $e\#d\#$ is less than $3M$.

7. The cases of $\vee$-IA and $\rightarrow$-IA are quite analogous.

8. In the case of $\rightarrow$-IS, we have

$$c_1 = A^*, \Gamma^* \Longrightarrow^* \Theta^*, B^*$$

$$d = \Gamma^* \Longrightarrow^* \Theta^*, (A \rightarrow B)^*.$$

Now, let $e_1 \stackrel{\text{def}}{=} \Longrightarrow^* A^*, (A \rightarrow B)^*$ and $e_2 \stackrel{\text{def}}{=} B^* \Longrightarrow^* (A \rightarrow B)^*$. By Lemma 3.5.13, the clauses e_1 and e_2 are subsumed by clauses in $\mathrm{def}_{\mathcal{D}}$. Now, let $e \stackrel{\text{def}}{=} \Gamma^* \Longrightarrow^* \Theta^*, B^*, (A \rightarrow B)^*$. Then e is subsumed by a resolvent of c_1 and e_1, and d is subsumed by a resolvent of e and e_2. So $e\#d\#$ is a derivation of d from S by resolution with subsumption. Exactly the same estimate for the complexities of this derivation holds as in the other cases.

9. In the case of $\forall$-IA, we have

$$c_1 = (F\{x\backslash t\})^*, \Gamma^* \Longrightarrow^* \Theta^*$$

$$d = (\forall x F)^*, \Gamma^* \Longrightarrow^* \Theta^*.$$

Now, let $e \stackrel{\text{def}}{=} (\forall x F)^* \Longrightarrow^* (F\{x\backslash t\})^*$. Then, by Lemma 3.5.13, the clause e is subsumed by a clause in $\text{def}_{\mathcal{D}}$. But d is subsumed by a resolvent of c_1 and e. So, $d\#$ is a derivation of d from S by resolution with subsumption. Its C-complexity is 1, its L-complexity is at most k, and its size is less than $3M$.

10. The case of $\exists$-IS is quite analogous.

11. In the case of $\forall$-IS, we have

$$c_1 = \Gamma^* \Longrightarrow^* \Theta^*, (F\{x\backslash a\})^*$$

$$d = \Gamma^* \Longrightarrow^* \Theta^*, (\forall x F)^*.$$

By Lemma 3.5.13, there exists a term T such that the clause $e \stackrel{\text{def}}{=} \{a \leftarrow T\}(F\{x\backslash a\})^* \Longrightarrow^* (\forall x F)^*$ is subsumed by some clause of $\text{def}_{\mathcal{D}}$. Since the eigenvariable a was assumed not to occur in the conclusion of the rule instance, it follows that it also does not occur in Γ^* and in Θ^*. So the clause $\Gamma^* \Longrightarrow^* \Theta^*, \{a \leftarrow T\}(F\{x\backslash a\})^*$ is an instance of the clause c_1. So the clause d is subsumed by a resolvent of the clauses c_1 and e. It follows that $d\#$ is a derivation of d from S by resolution with subsumption. Its C-complexity is 1, its L-complexity is at most k, and its size is less than $3M$.

12. The case of $\exists$-IA is quite analogous.

q.e.d.

3.6 Simulation of the Sequent Calculus in Extended Resolution

We have now all the major results that we need to prove the p-simulatability of the sequent calculus in extended definitional resolution. These are the Propositions 3.5.1, 3.5.3, 3.5.5, 3.5.4, and 3.2.1. We shall first consider the case of tree derivations. For tree derivations in the sequent calculus, the transforms of all occurrences of sequents are uniquely determined.

3.6.1 Tree Derivations

In this section we prove that a tree derivation in the sequent calculus can be p-simulated by a transformation to definitional form followed by a number of extensions and by a tree derivation in the resolution calculus, and that a cut free tree derivation in the sequent calculus can be p-simulated by a transformation to definitional form followed by a tree derivation in the resolution calculus.

Definition 3.6.1
Let S and T be sets of clauses, and let c be a clause. Let $\mathcal{E}$ be a derivation of T from S by extension only. Let $\mathcal{R}$ be a tree derivation of c from T in the resolution calculus. Then the derivation $\mathcal{ER}$ is said to be a *derivation* of c from S in *extended tree resolution*.

So, a derivation in extended tree resolution consists of a number of extension steps followed by a number of resolution and factorization steps. All input clauses and all the clauses obtained by extension steps can be used an arbitrary number of times as premises of resolution or factorization steps, but each other clause occurring in the derivation must not be used more than once as a premise. Note that the extension rule does not allow to generate again clauses it has already generated. This is due to the restriction on the predicate symbol and on the Skolem function it introduces that they must be new.

Definition 3.6.2
A *multiset* of clauses is a function from the set of clauses to the set of nonnegative integers. A clause c is said to be an *element* of a multiset M if $M(c) > 0$. We say that c is *contained* $M(c)$ *times* in M. If M_1 and M_2 are two multisets then the *union* $M_1 + M_2$ of M_1 and M_2 is the multiset defined by $(M_1 + M_2)(c) \stackrel{\text{def}}{=} M_1(c) + M_2(c)$. Similarly, $(M_1 - M_2)(c) \stackrel{\text{def}}{=} M_1(c) - M_2(c)$ if the so defined function $M_1 - M_2$ is a multiset. If c is a clause then we denote by $\{c\}$ the multiset defined by $\{c\}(c) \stackrel{\text{def}}{=} 1$, and $\{c\}(d) \stackrel{\text{def}}{=} 0$ for all clauses $d \neq c$. We shall use this notation if it is clear from the context that not a set but a multiset is meant by $\{c\}$. If $c_1, \ldots, c_k$ are clauses then we denote by $\{c_1, \ldots, c_k\}$ the multiset $\{c_1\} + \cdots + \{c_k\}$, provided again that it is clear from the context that we are referring to multisets. If M and N are multisets then $M \subseteq N$ means that $M(c) \leq N(c)$ for all clauses c.

Definition 3.6.3
Let S be a set of clauses, let M be a multiset of clauses, and let c be a clause. Let M' be the set of all elements of M. Let $\mathcal{Q}$ be a tree derivation of c from $S \cup M'$ by resolution with subsumption. Assume that each clause which is not in S occurs at most as many times as a leaf of $\mathcal{Q}$ as it is contained in the multiset M. Then we say that $\mathcal{Q}$ is a *tree derivation* of c from M *based upon* S.

Lemma 3.6.1
Let S be a set of clauses, let $M_1, \ldots, M_k$ and M be multisets of clauses, and let $c_1, \ldots, c_k$ and c be clauses such that $\{c_1, \ldots, c_k\} \subseteq M$. Let $\mathcal{Q}_j$ be a tree derivation of c_j from M_j based upon S for $j = 1, \ldots, k$, and let $\mathcal{Q}$ be a tree derivation of c from M based upon S. Then $\mathcal{Q}_1 \ldots \mathcal{Q}_k\mathcal{Q}$ is a tree derivation of c from $M_1 + \cdots + M_k + (M - \{c_1, \ldots, c_k\})$ based upon S.

Lemma 3.6.2
Under the assumptions of Proposition 3.5.3, the derivation $\mathcal{Q}$ constructed in the proof of Proposition 3.5.3 is a tree derivation of c from $\text{def}_{\mathcal{D}}$.

Proof: Under the assumptions made in Lemma 3.4.4, the following holds. For each clause $c \in \text{equ}[\sigma L_F, \tau L_G]$, the construction in the proof of Lemma 3.4.4 yields a tree derivation of c from a multiset $\{c_1, \ldots, c_k\}$ based upon the clause set $\text{def}_F \cup \text{def}_G$, where $c_j \in \text{equ}[\sigma L_{F_j}, \tau L_{G_j}]$. This is easily verified in each case but a bit tedious to write down in all detail and will not be explicated here any further. It follows from Lemma 3.6.1 by induction that, under the assumptions of Proposition 3.4.1, the derivation constructed in the proof of Proposition 3.4.1 for any clause $c \in \text{equ}[\sigma L_F, \tau L_G]$, is a tree derivation of c from the empty multiset $\{\}$ based upon the clause set $\text{def}_F \cup \text{def}_G$. In other words, the derivation is a tree derivation of c from $\text{def}_F \cup \text{def}_G$. So, under the assumptions of Proposition 3.5.2, for each clause $c \in \text{equ}[A, B]$, Proposition 3.5.2 yields a tree derivation of c from $\text{def}_{\mathcal{D}}$. So, under the assumptions of Proposition 3.5.3, the derivation of the clause c obtained by the construction in the proof of Proposition 3.5.3 is a tree derivation of c from $\text{def}_{\mathcal{D}}$.

q.e.d.

Lemma 3.6.3

Under the assumptions of Proposition 3.5.5, the derivation $\mathcal{Q}$ constructed in the proof of Proposition 3.5.5 is a tree derivation of the clause d from the multiset $\{c_1, \ldots, c_r\}$ based upon the clause set $\text{def}_{\mathcal{D}}$.

Proof: This is easily checked and trivial in each case.

q.e.d.

Proposition 3.6.1

Let $\mathcal{D}$ be a tree derivation of a formula F in the sequent calculus, let N be the size of $\mathcal{D}$, let n be the maximal structural complexity of formulas occurring in $\mathcal{D}$, and let k be the maximum of the cardinalities of the sequents of $\mathcal{D}$. Then there is a tree derivation $\mathcal{Q}$ of L_F from $\text{def}_{\mathcal{D}}$ *by resolution with subsumption such that the following inequalities hold.*

$$\begin{aligned} |\mathcal{Q}|_C &< 4N \\ |\mathcal{Q}|_L &< 9N \\ |\mathcal{Q}| &\leq 9Nn. \end{aligned}$$

Each clause of $\mathcal{Q}$ has at most $\max(3, k+1)$ *literals.*

Proof: Let $\mathcal{D} = (\mathcal{S}_1, \ldots, \mathcal{S}_r)$. Since $\mathcal{D}$ is a tree derivation, it has a tree expansion, and its sequents can be reordered according to the post order of the tree. The resulting derivation is still a tree derivation and has the same size, and the same maximal structural complexity of occurring formulas, as the derivation $\mathcal{D}$. So we can, without loss of generality, assume that $\mathcal{D}$ is in post order[10].

[10]The following construction works also without this assumption, but a stronger lemma than Lemma 3.6.1 is needed to prove this.

Let $c_1, \ldots, c_r$ be the transforms of the occurrences of $\mathcal{S}_1, \ldots, \mathcal{S}_r$, respectively. Since $\mathcal{D}$ is a tree derivation, these transforms are uniquely determined. We define derivations $\mathcal{Q}_1, \ldots \mathcal{Q}_r$ in resolution with subsumption as follows. Assume $j \leq r$. If $\mathcal{S}_j$ is an axiom then let $\mathcal{Q}_j$ be the derivation of c_j from $\mathrm{def}_{\mathcal{D}}$ as constructed in the proof of Proposition 3.5.3. If $\mathcal{S}_j$ is obtained from previous sequents $\mathcal{S}_{j_1}, \ldots, \mathcal{S}_{j_q}$ by a rule application then let $\mathcal{Q}_j$ be the derivation of c_j from $\mathrm{def}_{\mathcal{D}} \cup \{c_{j_1}, \ldots, c_{j_q}\}$ as constructed in the proof of Proposition 3.5.5. In the case of an axiom, Lemma 3.6.2 implies that the derivation $\mathcal{Q}_j$ is a tree derivation of c_j from $\mathrm{def}_{\mathcal{D}}$, i.e., a tree derivation of c_j from the empty multiset based upon the clause set $\mathrm{def}_{\mathcal{D}}$. In the case of a conclusion of a rule, Lemma 3.6.3 implies that the derivation is a tree derivation of c_j from the multiset $\{c_{j_1}, \ldots, c_{j_q}\}$ based upon $\mathrm{def}\mathcal{D}$.

Now, let $\mathcal{Q} \stackrel{\mathrm{def}}{=} \mathcal{Q}_1 \ldots \mathcal{Q}_r$. Then it follows from Lemma 3.6.1 by induction that $\mathcal{Q}$ is a tree derivation of $\{L_F\}$ from $\mathrm{def}_{\mathcal{D}}$ in resolution with subsumption.

Now, let N_j be the size of $\mathcal{S}_j\#$, and let n_j be the maximum of the structural complexities of the members of $\mathcal{S}_j$.

In the case of an axiom $H \Longrightarrow H$, the size of H is $\frac{1}{2}N_j - 1$, and we have $n_j < \frac{1}{2}N_j$ and $n_j \leq n$. So, by Proposition 3.5.3, we have $|\mathcal{Q}_j|_C \leq 8n_j < 4N_j$, and $|\mathcal{Q}_j|_L \leq 18m_j < 9N_j$, and $|\mathcal{Q}_j| \leq (16(\frac{1}{2}N_j - 1) + 24)n_j \leq 8(N_j + 1)n$. If $\mathcal{S}_j$ is an axiom then the word $\mathcal{S}_j\#$ cannot be shorter than the word $P() \Longrightarrow P()\#$ where P is a nullary predicate symbol. So its size is at least 8, and therefore, $1 \leq \frac{1}{8}N_j$. So we have $|\mathcal{Q}_j| \leq 9N_jn$.

In the case of a conclusion of a rule, Proposition 3.5.5 implies the same inequalities $|\mathcal{Q}_j|_C < 4N_j$, and $|\mathcal{Q}_j|_L < 9N_j$, and $|\mathcal{Q}_j| \leq 9N_jn$.

Since $N = \sum_{j=1}^{r} N_j$, and because of Lemma 3.2.2, the asserted inequalities follow.

q.e.d.

We shall need the following lemma which follows almost immediately from the definition of n-subsumption, Definition 2.3.5.

Lemma 3.6.4
Let P be an n-ary predicate symbol symbol and let $x_1, \ldots, x_n$ be pairwise distinct variables. Let c be the unit clause $\{Px_1 \ldots x_n\}$, and let d be a non-empty clause which n-subsumes the clause c. Then d is a variant of c.

Definition 3.6.4
A *derivation* of a formula F in *extended definitional tree resolution* is a derivation of $\{L_F\}$ from def_F in extended tree resolution.

Proposition 3.6.2
Let $\mathcal{D}$ be a tree derivation of a formula F in the sequent calculus, and let N be the size of $\mathcal{D}$. Then there is a derivation $\mathcal{X}$ of F in extended definitional tree resolution whose size is less than $45N^4$. If $\mathcal{D}$ is cut-free then $\mathcal{X}$ contains no extension steps.

Proof: From Proposition 3.6.1 it follows that there is a tree derivation $\mathcal{Q}$ of L_F from $\mathrm{def}_{\mathcal{D}}$ in resolution with subsumption, such that $|\mathcal{Q}| \leq 9N^2$ and such that the cardinalities of the clauses of $\mathcal{Q}$ are less than or equal to N. From Proposition 3.5.1 it follows that there is a derivation $\mathcal{E}$ of $\mathrm{def}_\mathcal{D}$ from def_F by extension whose size is less than or equal to $13N^2$. Moreover, the size of any word representation of $\mathrm{def}_\mathcal{D}$ is less than or equal to $13N^2$. From Proposition 3.2.1 it follows now that there is a tree resolution derivation $\mathcal{R}$ of a clause d from $\mathrm{def}_\mathcal{D}$ such that d n-subsumes $\{L_F\}$ and $|\mathcal{R}| \leq 44N^4$. Since any derivation $\mathcal{D}$ in the sequent calculus has at least size 8, we have $|\mathcal{E}| < N^4$. Now, $\mathcal{ER}$ is a derivation of d from def_F in extended tree resolution, and its size is less than $45N^4$. By Lemma 3.6.4, the clause d is a variant of $\{L_F\}$. Let $\mathcal{X}$ be the derivation of $\{L_F\}$ from $\mathrm{def}_\mathcal{D}$ in extended tree resolution which is obtained from $\mathcal{ER}$ by replacing the last clause d with the clause $\{L_F\}$. Then $\mathcal{X}$ is a derivation of F in extended definitional tree resolution whose size is less than $45N^4$. If the given derivation $\mathcal{D}$ is cut-free then $\mathrm{def}_\mathcal{D} = \mathrm{def}_F$, and there are no extension steps in the constructed derivation $\mathcal{X}$.

q.e.d.

From this follows immediately

Theorem 3.6.1
Extended definitional tree resolution can p-simulate tree derivations in the sequent calculus. Definitional tree resolution can p-simulate cut-free tree derivations in the sequent calculus.

3.6.2 Arbitrary Derivations

In this section we prove that extended definitional resolution can p-simulate the sequent calculus, and that definitional resolution can p-simulate the cut-free sequent calculus.

We have seen that, in general, there is not a uniquely determined transform of an occurrence of a sequent in arbitrary derivations of the sequent calculus. Also, the inheritence relations on transforms for different paths in a derivation are not compatible with each other. So the construction of a derivation we used in the case of tree derivations does not work here. Rather, we have to supplement it with derivations of one transform of an occurrence of a sequent from another transform of the same occurrence.

Lemma 3.6.5
Let $\mathcal{D}$ be a derivation in the sequent calculus, and let $R = (\chi_1, \ldots, \chi_r, \psi)$ be a rule application in $\mathcal{D}$. Let d be a transform of ψ, and let $c_1, \ldots, c_r$ be transforms of $\chi_1, \ldots, \chi_r$. Let k be the cardinality of the conclusion, i.e., the number of occurrences of formulas as members of ψ, and let N be the sum of the sizes of the sequents $\tilde{\chi}_1, \ldots, \tilde{\chi}_r$, and $\tilde{\psi}$. Then there is a derivation $\mathcal{Q}$ of the clause d from the clause set $\mathrm{def}_\mathcal{D} \cup \{c_1, \ldots, c_r\}$ by resolution with subsumption, such that the following inequalities hold.

$$\begin{aligned} |\mathcal{Q}|_C &\leq 8N \\ |\mathcal{Q}|_L &\leq 18N + 2(k+2)^2 \\ |\mathcal{Q}| &\leq 32N^2. \end{aligned}$$

Each clause of $\mathcal{Q}$ has at most $\max(3,l)$ *literals where l is the maximum of the cardinalities of $\tilde{\chi}_1,\ldots,\tilde{\chi}_r$ and $\tilde{\psi}$.*

Proof: Let $d_1,\ldots,d_r$ be the transforms of $\chi_1,\ldots,\chi_r$, respectively, which are inherited from the transform d of ψ. Let M_j be the size of the sequent $\tilde{\chi}_j$, and let k_j be the number of its members, for each $j = 1,\ldots,r$. Let M be the size of the sequent $\tilde{\psi}$, and let k be the number of its members. Then $N = M + \sum_{j=1}^{r} M_r$, and $k_j \leq k+1$. By Proposition 3.5.4 there is, for each $j = 1,\ldots,r$, a derivation $\mathcal{Q}_j$ of the clause d_j from the clauses set $\mathrm{def}_{\mathcal{D}} \cup \{c_j\}$ by resolution with subsumption such that

$$
\begin{aligned}
|\mathcal{Q}_j|_C &\leq 8M_j \\
|\mathcal{Q}_j|_L &\leq 18M_j + k_j^2 \\
|\mathcal{Q}_j| &\leq 32M_j^2.
\end{aligned}
$$

By Proposition 3.5.5, there is a derivation $\mathcal{Q}'$ of the clause d from the clause set $\mathrm{def}_{\mathcal{D}} \cup \{d_1,\ldots,d_r\}$ by resolution with subsumption such that

$$
\begin{aligned}
|\mathcal{Q}'|_C &\leq 2 \\
|\mathcal{Q}'|_L &\leq 2k+1 \\
|\mathcal{Q}'| &< 3M.
\end{aligned}
$$

Now, let $\mathcal{Q} \stackrel{\mathrm{def}}{=} \mathcal{Q}_1 \ldots \mathcal{Q}_r \mathcal{Q}'$. Then $\mathcal{Q}$ is a derivation of the clause d from the clause set $\mathrm{def}_{\mathcal{D}} \cup \{c_1,\ldots,c_r\}$ by resolution with subsumption. From Lemma 3.2.2 it follows that $|\mathcal{Q}|_C \leq 8\sum_{j=1}^{r} M_j + 2 \leq 8N$. Similarly, it follows that $|\mathcal{Q}|_L \leq 18\sum_{j=1}^{r} M_j + \sum_{j=1}^{r} k_j^2 + 2k + 1$. But this is less than or equal to $18N + r(k+1)^2 + 2k + 1$. Since the number r of premises of the clause is less than or equal to 2, we have $|\mathcal{Q}|_L \leq 18N + 2(k+2)^2$. The size of $\mathcal{Q}$ is by Lemma 3.2.2 less than $32\sum_{j=1}^{r} M_j^2 + 3M$ which is less than or equal to $32N^2$.

q.e.d.

Proposition 3.6.3
Let $\mathcal{D}$ be a derivation of a formula F in the sequent calculus. Let N be the size of $\mathcal{D}$. Then there is a derivation $\mathcal{Q}$ of L_F from $\mathrm{def}_{\mathcal{D}}$ *by resolution with subsumption such that the following inequalities hold.*

$$
\begin{aligned}
|\mathcal{Q}|_C &\leq 4N^2 \\
|\mathcal{Q}|_L &\leq 11N^2 \\
|\mathcal{Q}| &\leq 32N^3
\end{aligned}
$$

Each clause of $\mathcal{Q}$ has at most $\max(3,k)$ *literals where k is the maximum of the cardinalities of the sequents of $\mathcal{D}$.*

Proof: Let $\mathcal{D} = (\mathcal{S}_1, \ldots, \mathcal{S}_q)$. Let $\chi_1, \ldots, \chi_q$ be the respective occurrences of $\mathcal{S}_1, \ldots, \mathcal{S}_q$. Let us choose transforms $c_1, \ldots, c_q$ of $\chi_1, \ldots, \chi_q$, respectively. These transforms can be chosen in an arbitrary way. For $j = 1, \ldots, q$, let $\mathcal{Q}_j$ be defined as follows. If $\mathcal{S}_j$ is an axiom then let $\mathcal{Q}_j$ be the derivation of c_j from $\text{def}_{\mathcal{D}}$ by resolution with subsumption as constructed in Proposition 3.5.3. If $\mathcal{S}_j$ has been obtained from previous sequents $\mathcal{S}_{j_1}, \ldots, \mathcal{S}_{j_r}$ by an application of a rule of the sequent calculus then let $\mathcal{Q}_j$ be the derivation of the clause c_j from the clause set $\text{def}_{\mathcal{D}} \cup \{c_{j_1}, \ldots, c_{j_r}\}$ by resolution with subsumption as constructed in Lemma 3.6.5. Now, let k_j be the number of members of $\mathcal{S}_j$. Then, Proposition 3.5.3 and Lemma 3.6.5 imply that the following inequalities hold for all $j = 1, \ldots, q$.

$$\begin{aligned} |\mathcal{Q}_j|_C &\leq 8N \\ |\mathcal{Q}_j|_L &\leq 18N + 2(k_j + 2)^2 \\ |\mathcal{Q}_j| &\leq 32N^2. \end{aligned}$$

Now, let $\mathcal{Q} \stackrel{\text{def}}{=} \mathcal{Q}_1 \ldots \mathcal{Q}_q$. The number q of sequents in the derivation is at most $\frac{1}{2}N$. Furthermore, we have $\sum_{j=1}^{q}(k_j + 2)^2 \leq (\sum_{j=1}^{q}(k_j + 2))^2 \leq N^2$. The assertion follows now directly from Lemma 3.2.2.

q.e.d.

Proposition 3.6.4
Let $\mathcal{D}$ be a derivation of a formula F in the sequent calculus, and let N be the size of $\mathcal{D}$. Then there is a derivation $\mathcal{X}$ of F in extended definitional resolution whose size is less than $69N^5$. If $\mathcal{D}$ is cut-free then $\mathcal{X}$ contains no extension steps.

Proof: From Proposition 3.6.3 it follows that there is a derivation $\mathcal{Q}$ of L_F from $\text{def}_{\mathcal{D}}$ in resolution with subsumption, such that $|\mathcal{Q}| \leq 32N^3$ and such that the cardinalities of the clauses of $\mathcal{Q}$ are less than or equal to N. From Proposition 3.5.1 it follows that there is a derivation $\mathcal{E}$ of $\text{def}_{\mathcal{D}}$ from def_F by extension whose size is less than or equal to $13N^2$ which is in turn less than N^5 since the size of any derivation in the sequent calculus is greater than or equal to 8. Moreover, the size of any word representation of $\text{def}_{\mathcal{D}}$ is less than or equal to $13N^2$ which is in turn less than $2N^3$ since $N \geq 8$. From Proposition 3.2.1 it follows now that there is a resolution derivation $\mathcal{R}$ of a clause d from $\text{def}_{\mathcal{D}}$ such that d n-subsumes $\{L_F\}$ and $|\mathcal{R}| \leq 68N^5$. Now, $\mathcal{ER}$ is a derivation of d from def_F in extended resolution, and its size is less than $69N^5$. From Lemma 3.6.4 it follows that d is a variant of $\{L_F\}$. Let $\mathcal{X}$ be the derivation of $\{L_F\}$ from def_F which is obtained from $\mathcal{ER}$ by replacing its last clause d with the clause $\{L_F\}$. Then $\mathcal{X}$ is a derivation of F in extended definitional resolution whose size is less than $69N^5$. If the given derivation $\mathcal{D}$ is cut-free then $\text{def}_{\mathcal{D}} = \text{def}_F$, and there are no extension steps in the constructed derivation $\mathcal{X}$.

q.e.d.

So the following theorem holds.

Theorem 3.6.2
Extended definitional resolution can p-simulate the sequent calculus. Definitional resolution can p-simulate cut-free derivations in the sequent calculus.

3.7 Gentzen's Transformations

In [Gen35], Gerhard Gentzen has given transformations between derivations of a formula in a Frege-Hilbert calculus, in the natural deduction calculus, and in the sequent calculus. His transformations have polynomial time complexities and show the mutual p-simulatability of these three types of calculi. His transformation from the Frege-Hilbert calculus to natural deduction and his transformation from natural deduction to the sequent calculus map tree derivations to tree derivations. But his transformation from the sequent calculus to the Frege-Hilbert calculus does not have this property, and therefore it has to be changed slightly for our purposes. It then follows from the results of Section 3.6 that extended definitional resolution can p-simulate natural deduction and Frege-Hilbert calculi. Moreover, extended definitional tree resolution can p-simulate tree derivations in natural deduction and in Frege-Hilbert calculi.

We consider here three types of calculi, namely Frege-Hilbert calculi, the natural deduction calculus, and the sequent calculus. All these calculi have one thing in common. They allow to derive words over a certain alphabet. In the Frege-Hilbert calculus these words are formulas, in the natural deduction calculus they are judgements, and in the sequent calculus they are sequents. Each of these calculi consists of a number of *rules*. Each rule allows to derive one word, called the conclusion of the rule instance, from zero or more other words, called the premises of the rule instance. These rules can be considered as relations between finite sequences of words and words, in the sense indicated in the remark at the end of Section 2.1. A rule with no premises is called an *axiom*.

Definition 3.7.1
Let $\mathcal{A}$ be an alphabet, let $\mathcal{W}$ be a subset of the set of words over $\mathcal{A}$, and let $\mathcal{W}^*$ be the set of finite sequences of elements of $\mathcal{W}$. A *deduction rule* for (elements of) $\mathcal{W}$ is a subset of $\mathcal{W}^* \times \mathcal{W}$. An *instance* of a deduction rule R is an element of R. If $((w_1, \ldots, w_n), w)$ is an instance of a deduction rule then $w_1, \ldots, w_n$ are said to be the *premises* and w is said to be the *conclusion* of this instance.

Now assume that $\# \notin \mathcal{A}$.

Definition 3.7.2
Let R be a deduction rule for $\mathcal{W}$. Let $w_1, \ldots, w_n, w \in \mathcal{W}$. A *derivation* of w from $w_1, \ldots, w_n$ via R is a word $u_1\# \ldots \#u_r\#$ over the alphabet $\mathcal{A} \cup \{\#\}$ such that $u_r = w$ and such that, for each $k = 1, \ldots, r$, there is an instance of R whose premises are in $\{w_1, \ldots, w_n, u_1, \ldots, u_{k-1}\}$ and whose conclusion is u_k. A *strict tree derivation* of w from $w_1, \ldots, w_n$ via R is a derivation of w from $w_1, \ldots, w_n$ such that each occurrence of a word w_j in $w_1, \ldots, w_n$ and each occurrence of a word

u_j in $u_1\#\ldots\#u_r\#$ is used only once as a premise of an instance of a rule. By a *derivation* of a word $w \in \mathcal{W}$ via a deduction rule R we mean a derivation of w from the empty word via R.

We shall consider deduction rules for formulas, for judgements, and for sequents. For any given set $\mathcal{W}$ of words over a given alphabet, the union of any set of deduction rules is again a deduction rule. We shall sometimes speak of a derivation via several deduction rules. By this we mean a derivation via the union of these deduction rules.

Definition 3.7.3
Let R and R' be deduction rules. We say that R' is *polynomial* (or a *p-rule*) in (or relative to) R, if there is a polynomial p such that for any instance $((w_1,\ldots,w_n),w)$ of R' there is a derivation $\mathcal{D}$ of w from $w_1,\ldots,w_n$ in R whose size is less than or equal to $p(N)$ where

$$N \stackrel{\text{def}}{=} |w| + \sum_{u\in\{w_1,\ldots,w_n\}} |u|.$$

We say that R' is a *polynomial tree rule* (or a *pt-rule*) in (or relative to) R if, in addition, $\mathcal{D}$ is a strict tree derivation.

Note that, in this definition, N was not defined to be $|w| + \sum_{j=1}^{n} |w_n|$. If a formula occurs more than once as a premise of the rule instance, its size still occurs only once in the sum.

Lemma 3.7.1
Let $\mathcal{W}$ be a set of words over a given alphabet. Let R be a deduction rule for elements of $\mathcal{W}$. Then the union of any finite set of p-rules (pt-rules) relative to R, is again a p-rule (pt-rule) relative to R.

Lemma 3.7.2
If R' is a p-rule relative to a deduction rule R, and if R'' is a p-rule relative to R' then R'' is a p-rule relative to R. If R' is a pt-rule relative to a deduction rule R, and if R'' is a pt-rule relative to R' then R'' is a pt-rule relative to R.

Proof: Let p be the polynomial in the definition above for R and R', and let q be the polynomial for R' and R''. We can assume without loss of generality that p and q are monotonically increasing. Let $r \stackrel{\text{def}}{=} \lambda x.q(x)p(q(x))$. Then r is a polynomial. Now let $((w_1,\ldots,w_n),w)$ be an instance of R'', and let N be the sum of the sizes of the words $w_1,\ldots,w_n,w$. Then there is a derivation of w from $w_1,\ldots,w_n$ via R' whose size is less than or equal to $q(N)$. This derivation consists of at most $q(N)$ applications of the rule R', and for each of these rule applications, the sum of the sizes of the premises and of the conclusion is less than or equal to $q(N)$. Each rule application of R' can be converted to a derivation via R whose size is less than or equal to $p(q(N))$. Combining these derivations, we get a derivation of w from $w_1,\ldots,w_n$ via R whose size is less than or equal to $q(N)p(q(N))$, i.e., less than or equal to $r(N)$. Now, a combination of strict tree derivations yields again a strict tree derivation.

q.e.d.

We shall frequently denote deduction rules as follows.

$$\frac{W_1 \quad \dots \quad W_n}{W}$$

where $W_1, \dots, W_n$ and W are words over $\mathcal{A} \cup \mathcal{M}$, and $\mathcal{A}$ is the alphabet underlying the rule, and $\mathcal{M}$ is a set of meta variables (such as 'Γ', 'A', etc. in the sequent calculus). This is to be understood to denote the set of all pairs $((w_1, \dots, w_n), w)$ such that $w_1, \dots, w_n$ and w are obtained from $W_1, \dots, W_n$ and W by consistently replacing the metavariables with words, possibly subject to further restrictions stated in the formulation of the rule or elsewhere. If there are no premises in the rule we will just write W.

3.7.1 Simulation of the Frege-Hilbert Calculus in Natural Deduction

In this section we give a brief description of Gerhard Gentzen's transformation of derivations from the Frege-Hilbert calculus to the natural deduction calculus.

Definition 3.7.4
Let F be a formula. Then the *translation* F^* of F (to the natural deduction calculus) is the judgement $\Longrightarrow F$.

Now, each derivation of a formula F in the Frege-Hilbert calculus is transformed to a derivation of the translation of F in the natural deduction calculus. Each axiom of the Frege-Hilbert calculus is transformed to a derivation of its translation in the natural deduction calculus. Each rule application in the Frege-Hilbert calculus is transformed to a derivation of the translation of its conclusion from the translations of its premises in the natural deduction calculus.

For each axiom in the Frege-Hilbert calculus, Gentzen gives a derivation in the natural deduction calculus which is obviously a pt-rule. We only give two examples here, namely the derivations for A6 and A8.

$$\dfrac{\dfrac{\dfrac{\dfrac{\dfrac{A \Longrightarrow A \qquad A \to B \Longrightarrow A \to B}{A, A \to B \Longrightarrow B}\,(\to\text{-E}) \qquad B \to C \Longrightarrow B \to C}{A, A \to B, B \to C \Longrightarrow C}\,(\to\text{-E})}{A \to B, B \to C \Longrightarrow A \to C}\,(\to\text{-I})}{A \to B \Longrightarrow (B \to C) \to (A \to C)}\,(\to\text{-I})}{\Longrightarrow (A \to B) \to ((B \to C) \to (A \to C))}\,(\to\text{-I})$$

$$\dfrac{\dfrac{A \wedge B \Longrightarrow A \wedge B}{A \wedge B \Longrightarrow A}\,(\wedge\text{-E})}{\Longrightarrow A \wedge B \to A}\,(\to\text{-I})$$

The rule R1 of modus ponens in the Frege-Hilbert calculus is already a special case of the rule →-E of natural deduction. Gentzen's translation of the rule R2 of the Frege-Hilbert calculus to the natural deduction calculus is the derivation

$$\dfrac{\dfrac{\dfrac{A \Longrightarrow A \qquad \Longrightarrow A \rightarrow F\{x\backslash a\}}{A \Longrightarrow F\{x\backslash a\}}\ (\rightarrow\text{-E})}{A \Longrightarrow \forall x F}\ (\forall\text{-I})}{\Longrightarrow A \rightarrow \forall x F}\ (\rightarrow\text{-I})$$

and his translation of the rule R3 of the Frege-Hilbert calculus to the natural deduction calculus is the derivation

$$\dfrac{\dfrac{\exists x F \Longrightarrow \exists x F \qquad \dfrac{F\{x\backslash a\} \Longrightarrow F\{x\backslash a\} \qquad \Longrightarrow F\{x\backslash a\} \rightarrow A}{F\{x\backslash a\} \Longrightarrow A}\ (\rightarrow\text{-E})}{\exists x F \Longrightarrow A}\ (\exists\text{-E})}{\Longrightarrow \exists x F \rightarrow A}\ (\rightarrow\text{-I})$$

Definition 3.7.5
Let R be a rule of the Frege-Hilbert calculus. Then the *translation* R^* of R (to the natural deduction calculus) is the following rule for judgements.

$$\frac{A_1^* \quad \dots \quad A_n^*}{B^*}$$

whenever

$$\frac{A_1 \quad \dots \quad A_n}{B}$$

is an instance of the rule R in the Frege-Hilbert calculus.

Lemma 3.7.3
For each rule R of the Frege-Hilbert calculus, the rule R^* is a pt-rule relative to the natural deduction calculus.

Proof: If the rule of the Frege-Hilbert calculus is an axiom then Gentzen's transformation yields a derivation of the translation of that axiom in the natural deduction calculus. In each case it obvious by inspection that the size of the derivation is linear with respect to the size of the axiom. Moreover, in each case the derivation is a strict tree derivation. The same is true for the translations of the three rules.

q.e.d.

Proposition 3.7.1
The natural deduction calculus can p-simulate the Frege-Hilbert calculus. Tree derivations in the natural deduction calculus can be p-simulated by tree derivations in the Frege-Hilbert calculus.

Proof: By Lemma 3.7.3, there is a polynomial p such that for every instance $((F_1,\ldots,F_n),F)$ of a rule R of the Frege-Hilbert calculus, there is a derivation of F from $F_1,\ldots,F_n$ in the natural deduction calculus whose size is less than or equal to $p(N)$ where N is the sum of the sizes of $F_1^*,\ldots,F_n^*$, and F^*. Since the size of a transform F^* of a formula is the size of F plus 1, the size of the deduction can be bounded from above by a polynomial q of the sum of the sizes of $F_1,\ldots,F_n$, and F. Let $\mathcal{D}$ be a derivation of a formula F in the Frege-Hilbert calculus. Let N be the size of $\mathcal{D}$. Then there are at most N rule applications in $\mathcal{D}$, and for each rule application, the sum of the sizes of its premises and of its conclusion is less than or equal to N. Therefore, the total size of the derivation resulting from Gentzen's transformation, is less than or equal to $Nq(N)$. If the derivation $\mathcal{D}$ is a tree derivation then the derivation resulting from Gentzen's transformation is again a tree derivation since each of its constituent derivations (corresponding to the respective rule applications in $\mathcal{D}$) is a tree derivation.

q.e.d.

Remark It can be shown that the simulation is even linear.

3.7.2 Simulation of Natural Deduction in the Sequent Calculus

In this section we give a description of Gerhard Gentzen's transformation that he used in [Gen35] to show that natural deduction can be simulated in the sequent calculus. The issue of this section is to show that Gentzen's transformation yields a p-simulation.

We shall give a translation of judgements of natural deduction to sequents of the sequent calculus. Since in the sequent calculus[11] there is no falsum, $\perp$, we have to define it in terms of the other propositional connectives when we want to translate judgements to sequents. We choose an arbitrary nullary predicate symbol P. We shall use the symbol 'P' also to denote the formula $P()$ if it is clear from the context what we mean. Then $\perp$ will be translated to $P \wedge \neg P$.

Definition 3.7.6
Let F be a formula possibly containing the falsum $\perp$. Then the *translation* of F is the formula obtained from F by replacing all occurrences of $\perp$ with $P \wedge \neg P$. Let $A_1,\ldots,A_n$ and F be formulas possibly containing the falsum, and let $A_1^*,\ldots,A_n^*$, and F^* be the translations of $A_1,\ldots,A_n$, and F, respectively. Then the sequent $A_1^*,\ldots,A_n^* \Longrightarrow F^*$ is said to be a *translation* of the judgement $A_1,\ldots,A_n \Longrightarrow F$. If $\mathcal{J}$ is a judgement then we denote by $\mathcal{J}^*$ the translation of $\mathcal{J}$.

Definition 3.7.7
Let R be a rule of the natural deduction calculus. Then R^* denotes the following

[11] that is, in the version that we consider here — of course there are other versions of the sequent calculus that do contain the falsum —

rule for sequents.

$$\frac{\mathcal{J}_1^* \quad \cdots \quad \mathcal{J}_n^*}{\mathcal{J}^*}$$

if $\mathcal{J}$ is obtained from $\mathcal{J}_1, \ldots, \mathcal{J}_n$ by application of the rule R.

We shall prove that R^* is a pt-rule with respect to the sequent calculus for each rule R of the natural deduction calculus. From this it will follow that the sequent calculus can p-simulate natural deduction and that this also holds for tree derivations.

Definition 3.7.8
By the *weakening rule* we mean the following rule for sequents.

$$\frac{\Gamma \Longrightarrow F}{\Delta \Longrightarrow F}$$

where $\tilde{\Gamma} \subseteq \tilde{\Delta}$.

Lemma 3.7.4
The weakening rule is a pt-rule with respect to the sequent calculus.

Proof: Let $\mathcal{S}_1$ be the premise of an instance of the weakening rule, and let $\mathcal{S}_2$ be its conclusion. Then $\mathcal{S}_1$ has the form $\Gamma \Longrightarrow F$, and $\mathcal{S}_2$ has the form $\Delta \Longrightarrow F$ where $\tilde{\Gamma} \subseteq \tilde{\Delta}$. Let N be the sum of the sizes of $\mathcal{S}_1$ and $\mathcal{S}_2$. The number of occurrences of formulas as members of the antecedent of $\mathcal{S}_1$ is at most $\frac{1}{2}N$, and the same holds for $\mathcal{S}_2$. Now, we make the following sequence of deduction steps. We choose a formula $G \in \tilde{\Gamma}$ arbitrarily. We move all occurrences of G in the antecedent of $\mathcal{S}_1$ to the beginning of $\mathcal{S}_1$ by repeatedly applying the interchange rule of the sequent calculus. This can be done in at most $(\frac{N}{2} - m)m$ steps where m is the number of occurrences of G in $\mathcal{S}_1$. By at most $m-1$ contraction steps, we reduce the number of occurrences of G until it is less than or equal to the number of occurrences of G in $\mathcal{S}_2$. The number of derivation steps so far is less than or equal to $\frac{1}{2}Nm$. We do this in turn for each formula occurring in S_1. The total number of deduction steps so far is less than or equal to $\frac{1}{4}N^2$, and the sizes of the occurring sequents are less than N. Now we need at most $\frac{1}{2}N$ thinning steps to obtain a sequent that differs from $\mathcal{S}_2$ only with respect to the order of the formulas in the antecedent. With at most $\frac{1}{2}(\frac{N}{2}-1)\frac{N}{2}$ interchanges we obtain the sequent $\mathcal{S}_2$. The total number of steps is certainly less than N^2, and the sizes of the occurring sequents are less than N. So the total size of the derivation is less than N^3.

q.e.d.

Lemma 3.7.5
Let $\Gamma \Longrightarrow H$ be a judgement, and let N be its size. Then the size of the translation of $\Gamma \Longrightarrow H$ is less than or equal to $12N$.

Proof: The process of translation for a sequent consists in replacing one-character substrings $\bot$ with 12-character strings $(P() \wedge (\neg P()))$.

q.e.d.

Lemma 3.7.6
The zero-premise rule $\bot^* \Longrightarrow$ is a pt-rule relative to the sequent calculus.

Proof: Since this rule has only one instance, we only have to show that there is a derivation of $\bot^* \Longrightarrow$. Take the following derivation.

$$\cfrac{\cfrac{\cfrac{\cfrac{\cfrac{P \Longrightarrow P}{\neg P, P \Longrightarrow}\,(\neg\text{-IA})}{P \wedge \neg P, P \Longrightarrow}\,(\wedge\text{-IA})}{P, P \wedge \neg P \Longrightarrow}\,(\text{interchange})}{P \wedge \neg P, P \wedge \neg P \Longrightarrow}\,(\wedge\text{-IA})}{P \wedge \neg P \Longrightarrow}\,(\text{contraction})$$

q.e.d.

Lemma 3.7.7
For each rule R of the natural deduction calculus, the rule R^* is a pt-rule with respect to the rules of the sequent calculus.

Proof: For each instance of such a rule R^*, we give a strict tree derivation with the rules of the sequent calculus and the rules above which have been shown to be pt-rules with respect to the sequent calculus. It follows then from Lemma 3.7.2 that R^* is a pt-rule.

Assumption:

$$A \Longrightarrow A$$

Tertium non datur:

$$\cfrac{\cfrac{\cfrac{\cfrac{\cfrac{A \Longrightarrow A}{\Longrightarrow A, \neg A}\,(\neg\text{-IS})}{\Longrightarrow A, A \vee \neg A}\,(\vee\text{-IS})}{\Longrightarrow A \vee \neg A, A}\,(\text{interchange})}{\Longrightarrow A \vee \neg A, A \vee \neg A}\,(\vee\text{-IS})}{\Longrightarrow A \vee \neg A}\,(\text{contraction})$$

Ex falso quodlibet: Let $\tilde{\Delta} = \tilde{\Gamma}$.

$$\dfrac{\dfrac{\Gamma \Longrightarrow \bot^* \qquad \bot^* \Longrightarrow}{\Gamma \Longrightarrow}\ \text{(cut)}}{\Delta \Longrightarrow D}\ \text{(weakening)}$$

$\neg$-I: Let $\tilde{\Delta} = \tilde{\Gamma} \setminus \{A\}$.

$$\dfrac{\dfrac{\dfrac{\Gamma \Longrightarrow \bot^*}{A, \Delta \Longrightarrow \bot^*}\ \text{(weakening)} \qquad \bot^* \Longrightarrow}{A, \Delta \Longrightarrow}\ \text{(cut)}}{\Delta \Longrightarrow \neg A}\ (\neg\text{-IS})$$

$\neg$-E: Let $\tilde{\Theta} = \tilde{\Gamma} \cup \tilde{\Delta}$.

$$\dfrac{\dfrac{\Delta \Longrightarrow \neg A \qquad \dfrac{\Gamma \Longrightarrow A}{\neg A, \Gamma \Longrightarrow}\ (\neg\text{-IA})}{\Delta, \Gamma \Longrightarrow}\ \text{(cut)}}{\Theta \Longrightarrow \bot}\ \text{(weakening)}$$

$\wedge$-I: Let $\tilde{\Theta} = \tilde{\Gamma} \cup \tilde{\Delta}$.

$$\dfrac{\dfrac{\Gamma \Longrightarrow A}{\Theta \Longrightarrow A}\ \text{(weakening)} \qquad \dfrac{\Delta \Longrightarrow B}{\Theta \Longrightarrow B}\ \text{(weakening)}}{\Theta \Longrightarrow A \wedge B}\ (\wedge\text{-IS})$$

$\wedge$-E: Let $\tilde{\Delta} = \tilde{\Gamma}$.

$$\dfrac{\dfrac{\Gamma \Longrightarrow A \wedge B \qquad \dfrac{A \Longrightarrow A}{A \wedge B \Longrightarrow A}\ (\wedge\text{-IA})}{\Gamma \Longrightarrow A}\ \text{(cut)}}{\Delta \Longrightarrow A}\ \text{(weakening)}$$

$\vee$-I: Let $\tilde{\Delta} = \tilde{\Gamma}$.

$$\dfrac{\dfrac{\Gamma \Longrightarrow A}{\Gamma \Longrightarrow A \vee B}\ (\vee\text{-IS})}{\Delta \Longrightarrow A \vee B}\ \text{(weakening)}$$

$\vee$-E: Let $\tilde{\Lambda} = \tilde{\Gamma} \cup (\tilde{\Delta} \setminus \{A\}) \cup (\tilde{\Theta} \setminus \{B\})$.

$$\dfrac{\dfrac{\Gamma \Longrightarrow A \vee B \qquad \dfrac{\dfrac{\Delta \Longrightarrow C}{A, \Lambda \Longrightarrow C}\ \text{(weakening)} \qquad \dfrac{\Theta \Longrightarrow C}{B, \Lambda \Longrightarrow C}\ \text{(weakening)}}{A \vee B, \Lambda \Longrightarrow C}\ (\vee\text{-IA})}{\Gamma, \Lambda \Longrightarrow C}\ \text{(cut)}}{\Lambda \Longrightarrow C}\ \text{(weakening)}$$

$\rightarrow$-I: Let $\tilde{\Delta} = \tilde{\Gamma} \setminus \{A\}$.

$$\cfrac{\cfrac{\Gamma \Longrightarrow B}{A, \Delta \Longrightarrow B}\ \text{(weakening)}}{\Delta \Longrightarrow A \rightarrow B}\ (\rightarrow\text{-IS})$$

$\rightarrow$-E: Let $\tilde{\Theta} = \tilde{\Gamma} \cup \tilde{\Delta}$.

$$\cfrac{\cfrac{\Delta \Longrightarrow A \rightarrow B \qquad \cfrac{\Gamma \Longrightarrow A \qquad B \Longrightarrow B}{A \rightarrow B, \Gamma \Longrightarrow B}\ (\rightarrow\text{-IA})}{\Delta, \Gamma \Longrightarrow B}\ \text{(cut)}}{\Theta \Longrightarrow B}\ \text{(weakening)}$$

$\forall$-I: Let $\tilde{\Delta} = \tilde{\Gamma}$.

$$\cfrac{\cfrac{\Gamma \Longrightarrow F\{x\backslash a\}}{\Gamma \Longrightarrow \forall x F}\ (\forall\text{-IS})}{\Delta \Longrightarrow \forall x F}\ \text{(weakening)}$$

$\forall$-E: Let $\tilde{\Delta} = \tilde{\Gamma}$.

$$\cfrac{\cfrac{\Gamma \Longrightarrow \forall x F \qquad \cfrac{F\{x\backslash t\} \Longrightarrow F\{x\backslash t\}}{\forall x F \Longrightarrow F\{x\backslash t\}}\ (\forall\text{-IA})}{\Gamma \Longrightarrow F\{x\backslash t\}}\ \text{(cut)}}{\Delta \Longrightarrow F\{x\backslash t\}}\ \text{(weakening)}$$

$\exists$-I: Let $\tilde{\Delta} = \tilde{\Gamma}$.

$$\cfrac{\cfrac{\Gamma \Longrightarrow F\{x\backslash t\}}{\Gamma \Longrightarrow \exists x F}\ (\exists\text{-IS})}{\Delta \Longrightarrow \exists x F}\ \text{(weakening)}$$

$\exists$-E: Let $\tilde{\Theta} = \tilde{\Gamma} \cup (\tilde{\Delta} \setminus \{F\{x\backslash a\}\})$. Let Λ be the sequence of formulas obtained from Δ by removing from it all occurrences of $F\{x\backslash a\}$ as members of Δ.

$$\cfrac{\cfrac{\Gamma \Longrightarrow \exists x F \qquad \cfrac{\cfrac{\Delta \Longrightarrow C}{F\{x\backslash a\}, \Lambda \Longrightarrow C}\ \text{(weakening)}}{\exists x F, \Lambda \Longrightarrow C}\ (\exists\text{-IA})}{\Gamma, \Lambda \Longrightarrow C}\ \text{(cut)}}{\Theta \Longrightarrow C}\ \text{(weakening)} \quad .$$

q.e.d.

As a consequence, we have

Proposition 3.7.2
The sequent calculus can p-simulate the natural deduction calculus. Tree derivations in the natural deduction calculus can be p-simulated by tree derivations in the sequent calculus.

3.7.3 Simulation of the Sequent Calculus in the Frege-Hilbert Calculus

The transformation that Gentzen gave for the simulation of the sequent calculus in the Frege-Hilbert calculus is polynomial, but it does not map tree derivations to tree derivations. So we have to use a slightly different transformation here.

Let us choose an arbitrary but fixed nullary predicate symbol P. By $\top^*$ we denote the formula $P \rightarrow P$, and by $\bot^*$ we denote the formula $P \wedge \neg P$.

Definition 3.7.9
If Γ is a sequence of formulas separated from each other by commata then the formula Γ^- is defined by induction as follows.

1. If Γ is empty then $\Gamma^- \stackrel{\text{def}}{=} \top^*$.

2. $(A, \Gamma)^- \stackrel{\text{def}}{=} A \wedge \Gamma^-$.

Definition 3.7.10
If Γ is a sequence of formulas separated from each other by commata then the formula Γ^+ is defined by induction as follows.

1. If Γ is empty then $\Gamma^+ \stackrel{\text{def}}{=} \bot^*$.

2. $(A, \Gamma)^+ \stackrel{\text{def}}{=} \Gamma^+ \vee A$.

So, Γ^- is essentially the right associative conjunction of the members of Γ, and Γ^+ is essentially the left associative disjunction of the members of Γ.

Definition 3.7.11
Let $\mathcal{S} = \Gamma \Longrightarrow \Delta$ be a sequent. Then the *translation* $\mathcal{S}^*$ of $\mathcal{S}$ is the formula $\Gamma^- \rightarrow \Delta^+$.

Definition 3.7.12
A *conjunction* of a non-empty finite set of formulas is a formula inductively defined as follows. The formula A is a conjunction of the set $\{A\}$. If F is a conjunction of a set S of formulas and G is a conjunction of a set T of formulas then $(F \wedge G)$ is a conjunction of $S \cup T$. A *disjunction* of a set of formulas is defined analogously.

The *transitivity rule for* $\rightarrow$,

$$\frac{A \rightarrow B \qquad B \rightarrow C}{A \rightarrow C}$$

is a pt-rule relative to the Frege-Hilbert calculus because the conclusion can be obtained from the premises and from the axiom A6 by applying the modus ponens twice. The following zero-premise rules are also pt-rules relative to the Frege-Hilbert calculus as is easily verified.

$$A \to A \wedge A$$
$$A \wedge A \to A$$
$$A \wedge B \to B \wedge A$$
$$(A \wedge B) \wedge C \to A \wedge (B \wedge C)$$
$$A \wedge (B \wedge C) \to (A \wedge B) \wedge C$$
$$(A \to B) \to ((A \wedge C) \to (B \wedge C))$$
$$(A \to B) \to ((C \wedge A) \to (C \wedge B))$$

The first five of these implications state the idempotence, the commutativity, and associativity of $\wedge$ for formulas. They are directly applicable only at the outermost level of a formula. The last two of the above implications allow to extend the use of the idempotence, commutativity, and associativity laws to occurrences of $\wedge$ far inside a conjunction of a set of formulas. So, the following zero-premise rule is a pt-rule relative to the Frege-Hilbert calculus.

$$F \to G$$

where F is a (possibly nested) conjunction of formulas and G is obtained from F by replacing an occurrence of a conjunct $A \wedge A$ of F with A (or vice versa), by replacing an occurrence of a conjunct $A \wedge B$ of F with $B \wedge A$, or by replacing an occurrence of a conjunct $(A \wedge B) \wedge C$ of F with $A \wedge (B \wedge C)$ (or vice versa). The proof is by induction on the depth of the occurrence of that conjunction in the structure tree of the formula, and by giving an explicit estimate of the size of the derivation.

Definition 3.7.13
The $\wedge$-*reordering axiom scheme* is the following zero-premise rule.

$$F \to G \qquad \text{where } F \text{ and } G \text{ are conjunctions of a set } S.$$

Similarly, we have a $\vee$-reordering axiom scheme. Again by induction, the following lemma is obtained.

Lemma 3.7.8
The $\wedge$- and $\vee$-reordering axiom schemes are pt-rules relative to the Frege-Hilbert calculus.

Definition 3.7.14
The *reordering rule* is the following rule.

$$\frac{A \to B}{C \to D}$$

where A and C are conjunctions of a finite set S of formulas, and B and D are disjunctions of a finite set T of formulas.

Lemma 3.7.9
The reordering rule is a pt-rule relative to the Frege-Hilbert calculus.

Proof: By Lemma 3.7.8, the zero-premise rules $C \rightarrow A$ and $B \rightarrow D$ with the restrictions stated in the definition of the reordering rule, are pt-rules relative to the Frege-Hilbert caluclus. Since the transitivity rule for $\rightarrow$ is also a pt-rule, the assertion of the lemma follows.

q.e.d.

We introduce, for each rule of the sequent calculus, a corresponding rule for formulas (rather than sequents). We denote this rule by the name of the rule in the sequent calculus with an asterisk appended to it.

AXIOMS*

$$D \rightarrow D$$

Structural inferences:*

THINNING*

$$\frac{G \rightarrow T}{D \wedge G \rightarrow T} \qquad \frac{G \rightarrow T}{G \rightarrow T \vee D}$$

CONTRACTION*

$$\frac{D \wedge (D \wedge G) \rightarrow T}{D \wedge G \rightarrow T} \qquad \frac{G \rightarrow (T \vee D) \vee D}{G \rightarrow T \vee D}$$

INTERCHANGE*

$$\frac{(\Delta, D, E, \Gamma)^- \rightarrow T}{(\Delta, E, D, \Gamma)^- \rightarrow T} \qquad \frac{G \rightarrow (\Theta, E, D, \Lambda)^+}{G \rightarrow (\Theta, D, E, \Lambda)^+}$$

CUT*

$$\frac{\Gamma^- \rightarrow \Theta^+ \vee D \qquad D \wedge \Delta^- \rightarrow \Lambda^+}{(\Gamma, \Delta)^- \rightarrow (\Theta, \Lambda)^+}$$

Operational inferences:*
¬-introductions*:

¬-IA*

$$\frac{G \rightarrow T \vee A}{\neg A \wedge G \rightarrow T}$$

¬-IS*

$$\frac{A \wedge G \rightarrow T}{G \rightarrow T \vee \neg A}$$

∧-introductions*:

∧-IA*

$$\frac{A \wedge G \rightarrow T}{(A \wedge B) \wedge G \rightarrow T} \qquad \frac{B \wedge G \rightarrow T}{(A \wedge B) \wedge G \rightarrow T}$$

$\wedge$-IS*
$$\frac{G \to T \vee A \qquad G \to T \vee B}{G \to T \vee (A \wedge B)}$$

$\vee$-introductions*:

$\vee$-IA*
$$\frac{A \wedge G \to T \qquad B \wedge G \to T}{(A \vee B) \wedge G \to T}$$

$\vee$-IS*
$$\frac{G \to T \vee A}{G \to T \vee (A \vee B)} \qquad \frac{G \to T \vee B}{G \to T \vee (A \vee B)}$$

$\to$-introductions*:

$\to$-IA*
$$\frac{\Gamma^- \to \Theta^+ \vee A \qquad B \wedge \Delta^- \to \Lambda^+}{(A \to B) \wedge (\Gamma, \Delta)^- \to (\Theta, \Lambda)^+}$$

$\to$-IS*
$$\frac{A \wedge G \to T \vee B}{G \to T \vee (A \to B)}$$

$\forall$-introductions*:

$\forall$-IA*
$$\frac{F\{x\backslash t\} \wedge G \to T}{\forall x F \wedge G \to T}$$

$\forall$-IS*
$$\frac{G \to T \vee F\{x\backslash a\}}{G \to T \vee \forall x F} \qquad \text{where } a \text{ is eigenvariable}$$

$\exists$-introductions*:

$\exists$-IA*
$$\frac{F\{x\backslash a\} \wedge G \to T}{\exists x F \wedge G \to T} \qquad \text{where } a \text{ is eigenvariable}$$

$\exists$-IS*
$$\frac{G \to T \vee F\{x\backslash t\}}{G \to T \vee \exists x F}$$

There is the restriction on the critical rules $\forall$-IS* and $\exists$-IA* that the eigenvariable a must not occur in the conclusion.

Lemma 3.7.10
For every rule R of the sequent calculus, the rule R^* is a pt-rule relative to the Frege-Hilbert calculus.

Proof: The contraction* rule and the interchange* rule are subsets of the reordering rule. To see that the cut* rule is a pt-rule, note that the zero-premise rule

$$(G \to T \vee D) \to ((D \wedge Z \to L) \to (G \wedge Z \to T \vee L))$$

is a pt-rule, and that from this rule[12] and from the reordering rule and from modus ponens, the cut* rule is obtained by composition. Namely, from the first premise of the cut* rule and from the rule above we obtain by modus ponens (R1) the formula

$$(D \wedge \Delta^- \to \Lambda^+) \to (\Gamma^- \wedge \Delta^- \to \Theta^+ \vee \Lambda^+),$$

and further, from the second premise of the cut* rule together with this formula, again by modus ponens, the formula

$$\Gamma^- \wedge \Delta^- \to \Theta^+ \vee \Lambda^+,$$

and thus we obtain by the reordering rule the conclusion of cut*. Similarly, the rule $\to -IA^*$ is obtained from the zero-premise pt-rule

$$(G \to T \vee A) \to ((B \wedge Z \to L) \to ((A \to B) \wedge (G \wedge Z) \to T \vee L))$$

and from the reordering rule and from modus ponens by composition. For ∀-IS*, we have

$$\cfrac{\cfrac{\cfrac{G \to T \vee F\{x\backslash a\}}{G \wedge \neg T \to F\{x\backslash a\}}\ (\mathrm{I1})}{G \wedge \neg T \to \forall x F}\ (\mathrm{R2})}{G \to T \vee \forall x F}\ (\mathrm{I2})$$

where I1 is obtained by modus ponens from the pt-rule

$$(G \to T \vee H) \to (G \wedge \neg T \to H)$$

and I2 is obtained by modus ponens from the pt-rule

$$(G \wedge \neg T \to H) \to (G \to T \vee H).$$

The case of ∃-IA* is quite analogous. Now let us consider any of the rules Axioms*, Thinning*, ¬-IA*, ¬-IS*, ∧-IA*, ∧-IS*, ∨-IA*, ∨-IS*, →-IS*, ∀-IA*, or ∃-IS*. Let $A_1, \ldots, A_n$ be the premises of the rule, and let B is its conclusion. In all these cases

$$A_1 \to (A_2 \to (A_3 \to \ldots (A_n \to B)\ldots))$$

is easily seen to be a pt-rule, and with n applications of modus ponens we obtain the required rule.

q.e.d.

We have, as immediate consequences,

[12]with G instantiated to Γ^-, with T instantiated to Θ^+, with D instantiated to D, with Z instantiated to Δ^-, and with L instantiated to Λ^+

Lemma 3.7.11
The following rule is a pt-rule.

$$\frac{\mathcal{S}^*}{\mathcal{T}^*}$$

where $\mathcal{S}$ *and* $\mathcal{T}$ *are sequents such that* $\frac{\mathcal{S}}{\mathcal{T}}$ *is an instance of a rule of the sequent calculus.*

Proposition 3.7.3
The Frege-Hilbert calculus can p-simulate the sequent calculus. Tree derivations in the sequent calculus can be p-simulated by tree derivations in the Frege-Hilbert calculus.

We can combine the results of this section in the following theorem.

Theorem 3.7.1
The sequent calculus, the natural deduction calculus, and the Frege-Hilbert calculus can p-simulate each other. The same holds for the restrictions of these three calculi to tree derivations.

3.8 Definitions

One feature used frequently in every day mathematics, and also in the use of natural language, is the construct of definitions. Gottlob Frege introduced this feature already in his Begriffsschrift [Fre79]. It has been recognized early that definitions can be eliminated from a language of logic. This may, however, lead to a great increase of the sizes of formulas, and also of the sizes of their proofs.

We have seen that the extension rule can be viewed as a way to make definitions in clausal form logic. In this section we shall show how a definition in full first order logic can be transformed to a number of extension steps in clausal form logic. The transformation is in accordance with the transformation of a formula to definitional form.

The claim of this section is that such calculi as Frege-Hilbert calculi, natural deduction, or the sequent calculus, can be p-simulated in extended definitional resolution even when they are augmented by the feature of definition.

Definition 3.8.1
Let $\mathcal{L}$ be a language of first order logic. Then a *definition* relative to $\mathcal{L}$ is a formula of the form

$$\forall x_1 \ldots \forall x_k (P(x_1, \ldots, x_k) \leftrightarrow F)$$

where

1. F is a formula of $\mathcal{L}$.

2. $x_1, \ldots, x_k$ are pairwise distinct variables, and all variables occurring free in F are in the set $\{x_1, \ldots, x_k\}$.

3. P is a k-ary predicate symbol not in $\mathcal{L}$.

The predicate symbol P is said to be the predicate symbol *defined* in the definition F.

Definition 3.8.2
Let $\mathcal{L}$ be a language of first order logic. Then a *definition sequence* relative to $\mathcal{L}$ is a finite sequence $(F_1, \ldots, F_n)$ of formulas such that each F_m is a definition relative to the language whose symbols are the symbols of $\mathcal{L}$ and the symbols defined in the definitions $F_1, \ldots, F_{m-1}$. The set $\{F_1, \ldots, F_n\}$ is said to be a *definition set* relative to $\mathcal{L}$.

Proposition 3.8.1
Let S be a set of formulas in a language $\mathcal{L}$, and let F be a formula in $\mathcal{L}$. Let T be a definition set relative to the language $\mathcal{L}$. Then

$$S \models F \iff S \cup T \models F.$$

Proof: The direction from left to right is trivial. So let $S \cup T \models F$, and let ι be a model of S. We have to prove that $\iota(F)$ is true. Since T is a definition set relative to $\mathcal{L}$, there is a definition sequence $(F_1, \ldots, F_n)$ relative to $\mathcal{L}$ such that $T = \{F_1, \ldots, F_n\}$. Let $P_1, \ldots, P_n$ be the predicate symbols defined in $F_1, \ldots, F_n$, respectively. For $j = 1, \ldots, n$, let $\mathcal{L}_j$ be the language whose symbols are the symbols of $\mathcal{L}$ and the predicate symbols $P_1, \ldots, P_j$. Let $\iota_0 \stackrel{\text{def}}{=} \iota$, and for $j = 1, \ldots, n$, let ι_j be the extension of the interpretation ι_{j-1} to the language $\mathcal{L}_j$ such that $\iota_j(F_j)$ is true. Then ι_n is a model of T. Since ι is a model of S and since ι and ι_n coincide on $\mathcal{L}$, it follows that ι_n is a model of S, and therefore also a model of $S \cup T$. Since $S \cup T \models F$, the interpretation ι_n must also be a model of F. Again, since ι and ι_n coincide on $\mathcal{L}$, it follows that $\iota(F)$ is true.

q.e.d.

An example of a definition set relative to a given language $\mathcal{L}$ is any set of formulas of the form D_G as defined in Section 2.2, where G is a formula of $\mathcal{L}$ and the predicate symbol P_G is not a symbol of $\mathcal{L}$. Let $\mathcal{L}$ be a first order language, and let F be a formula of $\mathcal{L}$. Let us assume that all the predicate symbols P_G such that G is a subformula of F, do not occur in the language $\mathcal{L}$. Then def_F is (the translation to a set of clauses of) a definition set relative to the language $\mathcal{L}$. Then Proposition 3.8.1 states that F is valid if and only if $\text{def}_F \models F$. But we also have $\text{def}_F \models F \leftrightarrow L_F$. So we can get a proof of the equivalence of $\models F$ and $\text{def}_F \models L_F$ from this proposition.

In such formal languages as the Begriffsschrift, there is an explicit language construct to introduce definitions. For a formula

$$\forall x_1 \ldots \forall x_k (P(x_1, \ldots, x_k) \leftrightarrow F)$$

that has been added to a set of formulas as a definition, we shall write

$$P(x_1, \ldots, x_k) \stackrel{\text{def}}{\leftrightarrow} F.$$

The introduction of definitions is justified by Proposition 3.8.1.[13]

In the following treatment of definitions, we consider $A \leftrightarrow B$ as shorthand for $(A \to B) \wedge (B \to A)$.

Definition 3.8.3
By a *derivation* of a formula F in the *sequent calculus with definitions*, in the *natural deduction calculus with definitions*, or in the *Frege-Hilbert calculus with definitions*, we mean a finite sequence $(\mathcal{T}_1, \ldots, \mathcal{T}_n)$ of sequents, or judgements, or formulas, respectively, such that the following conditions hold.

1. $\mathcal{T}_n$ is the sequent or judgement $\Longrightarrow F$, or the formula F, respectively.

2. For each $\mathcal{T}_j$, one of the following conditions holds.

   (a) $\mathcal{T}_j$ is a sequent or judgement $\Longrightarrow G$ or a formula G, where G is a definition relative to the language whose predicate and function symbols are the predicate and function symbols occurring in $\mathcal{T}_1, \ldots, \mathcal{T}_{j-1}, F$.

   (b) $\mathcal{T}_j$ is obtained from $\mathcal{T}_1, \ldots, \mathcal{T}_{j-1}$ by applying a rule[14] of the sequent calculus, the natural deduction calculus, or the Frege-Hilbert calculus, respectively.

So, effectively, definitions in any of these calculi are treated as a new kind of axioms with certain restrictions posed on the introduction of these axioms in a derivation.

We shall show how definitions in the sequent calculus, in natural deduction, or in the Frege-Hilbert calculus, can be simulated in extended definitional resolution. In order to prove results on the simulation of a derivation from such axioms, we have to introduce the following concepts.

Let S be a set of formulas. Then the *definitional form* Def_S of S is a set of clauses defined as follows

$$\mathrm{Def}_S \stackrel{\mathrm{def}}{=} \bigcup_{G \in S} \mathrm{def}_G \cup \{\{L_G\} \mid G \in S\}.$$

[13]Note, however, that it is essential in order for this proposition to hold, that a clear distinction be made between predicate symbols and terms. The difficulties that could arise if expressions such as $P(P)$ are allowed in a language, were pointed out by Bertrand Russel in a letter to Gottlob Frege (Engl. transl. in [Hei67]). *Russel's paradox* refers to the following "definition" which can be made in the Begriffsschrift.

$$P(x) \stackrel{\mathrm{def}}{\leftrightarrow} \neg x(x).$$

The formula corresponding to this definition is $\forall x(P(x) \leftrightarrow \neg x(x))$ which is a wellformed formula in the Begriffsschrift. This formula is inconsistent, and so the Begriffsschrift is inconsistent. The Begriffsschrift becomes consistent, however, if either the construction of definitions is disallowed, or if a suitable discipline for the application of predicate and function symbols is imposed by introducing a typed logic. The latter has been done by Alfred N. Whitehead and Bertrand Russel in their Principia Mathematica [WR13]. A particularly simple typed logic is first order predicate logic.

[14]Note that we consider an axiom of a calculus as a rule with no premises.

Lemma 3.8.1
A set S of formulas is satisfiable if and only if its definitional form is satisfiable. Let S be a set of formulas, and let F be a formula. Then

$$S \models F \iff \mathrm{def}_F \cup \mathrm{Def}_S \models L_F.$$

Proof: Let $\mathcal{L}$ be the language whose function symbols and predicate symbols are the function and predicate symbols occurring in the formulas of S or of $S \cup \{F\}$, respectively. Let $U \stackrel{\mathrm{def}}{=} \bigcup_{G \in S} \mathrm{def}_G$. Let $C \stackrel{\mathrm{def}}{=} \{\{L_G\} \mid G \in S\}$.

The clause set U is the translation of a definition set relative to $\mathcal{L}$, to a set of clauses. So, by Proposition 3.8.1, it follows that S is satisfiable iff $S \cup U$ is satisfiable. But $U \models G \leftrightarrow L_G$ for all $G \in S$. So, $S \cup U$ is satisfiable iff $U \cup C$ is satisfiable. But $U \cup C = \mathrm{Def}_S$. So we have proved the first part of the lemma.

Let $T \stackrel{\mathrm{def}}{=} \mathrm{def}_F \cup U$. Then T is the translation of a definition set relative to $\mathcal{L}$, to a set of clauses. From Proposition 3.8.1 it follows that $S \models F$ if and only if $S \cup T \models F$. But $T \models F \leftrightarrow L_F$, and $T \models G \leftrightarrow L_G$ for all $G \in S$. So, $S \cup T \models F$ iff $C \cup T \models L_F$. But $C \cup T = \mathrm{def}_F \cup \mathrm{Def}_S$. So, $S \models F$ iff $\mathrm{def}_F \cup \mathrm{Def}_S \models L_F$.

q.e.d.

Definition 3.8.4
A *derivation* of a formula F from a set S of formulas *in extended definitional resolution* is a derivation of the clause $\{L_F\}$ from the clause set $\mathrm{def}_F \cup \mathrm{Def}_S$ in extended resolution.

The results on simulation of the sequent calculus, natural deduction, and the Frege-Hilbert calculus, in extended definitional resolution, carry over to derivations of a formula from a set of formulas.

Proposition 3.8.2
Any derivation $\mathcal{D}$ of a formula F from a set S of formulas in the sequent calculus, in the natural deduction calculus, or in the Frege-Hilbert calculus can be transformed to a derivation of F from S in extended definitional resolution. There is a polynomial p such that, for all such derivations $\mathcal{D}$, the size of the transform of $\mathcal{D}$ is less than or equal to $p(|\mathcal{D}|)$.

Proof: Let us consider the case of the sequent calculus first. In contrast to the derivations we considered in the previous sections, in general not all of our axioms are of the form $H \Longrightarrow H$, but there may also be axioms of the form $\Longrightarrow H$ where $H \in S$. The concepts of descendants, subordinates, transforms, the clause set $\mathrm{def}_\mathcal{D}$, etc., are defined the same way as before.

The first difference is in Proposition 3.5.3. There, $\mathrm{def}_\mathcal{D}$ has to be replaced by $\mathrm{def}_\mathcal{D} \cup \mathrm{Def}_S$. In the proof of Proposition 3.5.3, we have to consider not only axioms of the form $H \Longrightarrow H$ now, but also axioms of the form $\Longrightarrow H$ with $H \in S$. So, let $\mathcal{S} = \Longrightarrow H$ with $H \in S$, let χ be an occurrence of $\mathcal{S}$ in $\mathcal{D}$, and let c be a transform of χ. We have to prove that, in resolution with subsumption, there is a derivation $\mathcal{R}$ of c, and that the inequalities stated in Proposition 3.5.3 hold for $\mathcal{R}$. The transform c of $\mathcal{S}$ has the form σL_G where $H \equiv \sigma G$. By Proposition 3.4.1, there is a derivation of equ$[L_H, \sigma L_G]$ from def$_H \cup$def$_G$ by resolution with subsumption which satisfies the required inequalities. In this derivation we delete the clause $L_H \vee \neg\sigma L_G$ (because we do not need this clause). Instead, we append the resolvent σL_G of the two clauses $\neg L_H \vee \sigma L_G$ and L_H. The latter clause is an element of Def$_S$. The resulting derivation is a derivation of σL_G, i.e., of c, from def$_\mathcal{D} \cup$ Def$_S$, and it still obeys the inequalities of Proposition 3.4.1. It follows that Proposition 3.5.3 holds also for derivations from a set S of formulas, if def$_\mathcal{D}$ is replaced with def$_\mathcal{D} \cup$ Def$_S$.

All the other lemmata and propositions proved in Sections 3.3, 3.5, and 3.6.2, and their proofs, are exactly the same for derivations from a set S of formulas in the sequent calculus, as they are for derivations from the empty set. Of course, def$_\mathcal{D}$ has to be replaced with def$_\mathcal{D} \cup$ Def$_S$. We then obtain from Proposition 3.6.4 the assertion.

Now, consider the cases of natural deduction and Frege-Hilbert calculus. We use Gentzen's transformations again to transform $\mathcal{D}$ to a derivation of F from S in the sequent calculus. Again, all the proofs directly carry over from Section 3.7. Since these transformations are polynomial, the resulting transformation from any one of these two calculi to extended definitional resolution must also be polynomial.

q.e.d.

Theorem 3.8.1
Extended definitional resolution can p-simulate the Frege-Hilbert calculus with definitions, the natural deduction calculus with definitions, and the sequent calculus with definitions.

Proof: Let $\mathcal{D}$ be a derivation of a formula F in the Frege-Hilbert calculus with definitions, in the natural deduction calculus with definitions, or in the sequent calculus with definitions. Without restriction of generality we can assume that all definitions in the given derivation preceed all other inference steps. So the derivation starts with a definition sequence $(D_1, \ldots, D_n)$ after which no more definition steps occur. Each D_j is the universal closure of a formula $L_j \leftrightarrow G_j$. Let P_j be the predicate symbol of L_j.

Now we simulate the definition sequence by the following extension steps. First we make all extensions generating the clauses in def$_{G_1}$. Then we make one extension step introducing P_1 as a new predicate symbol and adding the clauses $\neg L_1 \vee L_{G_1}$ and $L_1 \vee \neg L_{G_1}$. Then we make the extensions generating the clauses in def$_{G_2}$. Then we make one extension step introducing P_2 as a new predicate symbol and adding the clauses $\neg L_2 \vee L_{G_2}$ and $L_2 \vee \neg L_{G_2}$. We go on like this until we have generated the clauses of def$_{G_n}$ and introduced P_n as a new predicate symbol. Finally, we make all extension steps corresponding to def$_H$ where H is a D_j or a cut formula that has not yet been used for an extension step. Let us denote by $\text{def}_{\mathcal{D}}$ the set of clauses obtained from def_F through all these extension steps.

Now, each of the unit clauses $\{L_{D_j}\}$ has a short (6 resolution steps) derivation from the clause set $\text{def}_{\mathcal{D}}$ as is easily verified. Let

$$D \stackrel{\text{def}}{=} \text{def}_F \cup \bigcup_{j=1}^{n} \text{def}_{D_j}.$$

By Proposition 3.8.2, the derivation $\mathcal{D}$ can be transformed to a derivation of F from $\{D_1, \ldots, D_n\}$ in extended definitional resolution. By definition, this is a derivation of the clause $\{L_F\}$ from the clause set $D \cup \{\{L_{D_j}\} \mid j = 1, \ldots, n\}$. Since $D \subseteq \text{def}_{\mathcal{D}}$, and since there is a short derivation of each $\{L_{D_j}\}$ from $\text{def}_{\mathcal{D}}$, we obtain a resolution derivation of $\{L_F\}$ from $\text{def}_{\mathcal{D}}$. Together with the extension derivation of $\text{def}_{\mathcal{D}}$ from def_F, we obtain a derivation of $\{L_F\}$ from def_F in extended resolution. This derivation is a derivation of F in extended definitional resolution, and its size is bounded polynomially with respect to the size of $\mathcal{D}$.

q.e.d.

Chapter 4

Connection Structures

In Sections 2.4 and 2.5 we have shown that resolution can p-simulate (and even simulate step by step) the connection calculus, but that the connection calculus cannot p-simulate resolution. We have noted at the end of Section 2.5 that there are two main features which resolution has but the connection calculus does not have, and this is the main reason why the connection calculus cannot p-simulate resolution. One of these features is the use of lemmata. The other is the ability to forget variables.

In this chapter, we present a calculus — the connection structure calculus — which has both of these features but is based on the idea of the connection method. At the same time, this calculus is very close to resolution, and can be regarded as a generalized resolution with an additional structure on top of it (just as, for example, connection graph resolution can be regarded as resolution with an additional structure on top of it). The connection structure calculus can simulate step by step the connection calculus as well as the resolution calculus. Most of the results of this chapter have been published in [Ede89].

We shall see in Section 4.2 how a resolution refutation can be transformed to a connection proof at exponential cost. Each resolvent that is used more than once as a parent clause in the resolution refutation, corresponds to a connected submatrix in the connection proof, of which more than one variant is generated. A nested use of lemmata in resolution gives rise to an exponential number of variants of such a connected submatrix in the corresponding connection proof. Now, the connection structure calculus allows to code connection proofs in such a way that such connected submatrices are generated only once, and so any resolution refutation can be transformed to a derivation in the connection structure calculus at polynomial cost. In fact, there is a step by step correspondence between the resolution refutation and its transform.

4.1 Unifier Sets

In this section we build the concept of forgetting variables into the concept of a unifier. When we use unification we are primarily interested to get some handy description of the set of unifiers of some set of terms, or of the set of all (simultaneous) unifiers of some set of pairs of terms or literals. Such a handy description is usually yielded by giving a most general unifier. The idea of the concept of a unifier set defined below, is to describe the result of first obtaining the most general unifier of some set P, and then forgetting all variables except those in some specified set ξ.

Definition 4.1.1
Let P be a finite set of pairs of terms or of pairs of literals, and let ξ be a subset of the set $\mathcal{V}$ of variables. Then the *unifier set* $U[P,\xi]$ of P with respect to ξ is defined as the set of all restrictions of unifiers of P to the set ξ. If ξ is a finite set of variables then U_ξ denotes the set of all $U[P,\xi]$.

Obviously, if P is unifiable, then the set $U[P,\mathcal{V}]$ is the set of all unifiers of P, which can be described as the set of all substitutions that are more general than mgu(P). However, such a simple description is not always possible for $U[P,\xi]$. Let, for example, $P = \{(x, f(y,z)\}$ and $\xi = \{x\}$. Then

$$U[P,\xi] = \{\{x \leftarrow f(s,t)\} \mid s,t \text{ are terms}\}.$$

There is no single one among these substitutions that is more general than all of them. One way to get around this problem is to introduce, instead of y and z, two new "dummy" variables $y^\star$ and $z^\star$ which do not belong to $\mathcal{V}$. Then $\{x \leftarrow f(y^\star, z^\star)\}$ is, in fact, more general than all substitutions of $U[P,\xi]$. So we have a finite representation of $U[P,\xi]$.

Another way of arriving at a finite representation is by means of a dag. Let, for example,

$$P = \{\{x, f(u, g(v))\}, \{u, v\}, \{v, f(f(y, w), g(z))\}, \{g(w), g(z)\}\}$$

and $\xi = \{u, v, w, x, z\}$. To the unifier set of P corresponds the dag shown in Figure 4.1.

Each node that is not a leaf is marked by a function symbol whose arity equals the outdegree of the node. In addition, some of the leaves of the dag may be marked by constants. In Figure 4.1 there are three nodes marked with f, two nodes marked with g, and two nodes without marks. Moreover, each variable of ξ is associated with one and only one node of the dag. In Figure 4.1 the variables associated to each node are written to the left of the node separated from it by a $\sim$ character.

If more than one variable is associated with one node then those variables are shared variables. If ξ is the set of all variables occurring in P then each leaf which is not marked by a constant is associated with a variable. We have then just the natural dag representation for the most general unifier. If ξ is a proper subset of the set of all variables occurring in P then the variables not in ξ are just deleted in the dag representation, possibly leaving leaves which are not associated with variables. This is the case with the variable y and the node at the bottom left in

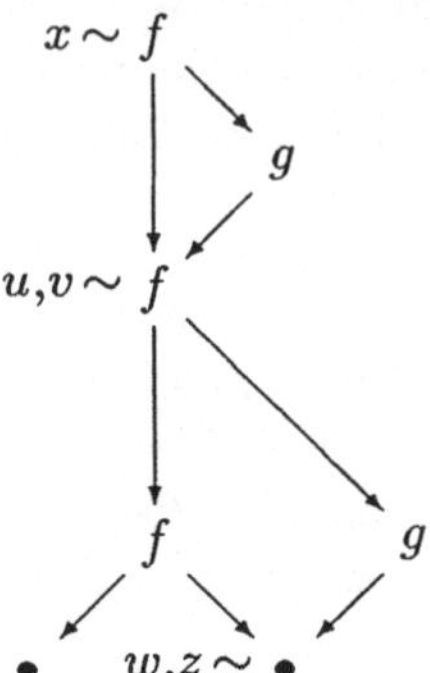

Figure 4.1: The dag representation of a unifier set

our example. On the other hand, if we choose $\xi = \{u, v, w, z\}$ in our example, then the root of the dag is not associated with a variable any more. It can be removed by a suitable garbage collection algorithm. Generally, any node can be removed that does not have an ancestor to which a variable is associated.

Similarly to the reduction of a dag by garbage collection, a reduction can be defined that involves merging two nodes marked by the same function symbol into one node if the two nodes have the same sons. Both kinds of reductions do not change the unifier set represented by the dag. This reducibility relation has then the property that two irreducible dags representing the same unifier set are always isomorphic. So there is a one-to-one correspondence between unifier sets and irreducible dags modulo isomorphism. We shall not go into any details of the dag representation of unifier sets here, but we shall define three basic operations defined on unifier sets:

Definition 4.1.2
Let ξ and η be finite sets of variables, let $\Gamma \in U_\xi$ and $\Delta \in U_\eta$, let $x \in \xi$ and let $f : \xi \to \mathcal{V}$ be an injection from ξ to the set $\mathcal{V}$ of variables. Then we define

- merge(Γ, Δ) as the set of all substitutions σ defined on the set of variables $\xi \cup \eta$ such that $\sigma|_\xi \in \Gamma$ and $\sigma|_\eta \in \Delta$.
- forget$(\Gamma, x) \stackrel{\text{def}}{=} \{\sigma|_{\xi \setminus \{x\}} \mid \sigma \in \Gamma\}$.
- rename$(\Gamma, f) \stackrel{\text{def}}{=} \{\sigma f^{-1} \mid \sigma \in \Gamma\}$.

The operation merge is quite analogous to the merge operation for ordinary unification. The forget operation corresponds to forgetting one variable, and the rename operation corresponds to replacing each variable x in the most general unifier by the variable $f(x)$. We shall show that each of these operations yields again a unifier set. To this end we need a few lemmata.

Lemma 4.1.1
Let P be a finite set of pairs of terms or a finite set of pairs of literals. Let ξ be a set of variables, and let π be a permutation of variables such that the domain of π and the set ξ are disjoint. Then $U[P,\xi] = U[\pi P,\xi]$.

Proof: The restriction to ξ of a unifier σ of P is identical to the restriction to ξ of the unifier $\sigma\pi^{-1}$ of πP, and the restriction to ξ of a unifier σ of πP is identical to the restriction to ξ of the unifier $\sigma\pi$ of P.

q.e.d.

Lemma 4.1.2
Let P and Q be finite sets of pairs of terms or finite sets of pairs of literals. Let $\mathcal{P}$ be the set of variables occurring in P, and let $\mathcal{Q}$ be the set of variables occurring in Q. Let ξ and η be finite sets of variables. Assume that $(\mathcal{P}\cup\xi)\cap(\mathcal{Q}\cup\eta) = \xi\cap\eta$. Then

$$\mathrm{merge}(U[P,\xi], U[Q,\eta]) = U[P\cup Q, \xi\cup\eta].$$

Proof: In order to see that the right hand side is a subset of the left hand side of the asserted equation, assume that σ is an element of the right hand side. Then σ is the restriction of a unifier τ of $P\cup Q$ to the set $\xi\cup\eta$. So, τ is a unifier of P and a unifier of Q. So, $\tau|_\xi \in U[P,\xi]$ and $\tau|_\eta \in U[Q,\eta]$. So, $\tau|_{\xi\cup\eta} \in \mathrm{merge}(U[P,\xi], U[Q,\eta])$, i.e., σ is an element of the left hand side of the asserted equation.

Now, let σ be an element of the left hand side of the asserted equation. Then $\sigma|_\xi \in U[P,\xi]$ and $\sigma|_\eta \in U[Q,\eta]$. So there is a unifier ρ of P such that $\sigma|_\xi = \rho|_\xi$, and there is a unifier τ of Q such that $\sigma|_\eta = \tau|_\eta$. It follows that ρ and τ coincide with each other on the set $\xi\cap\eta$ which is by assumption equal to the set $(\mathcal{P}\cup\xi)\cap(\mathcal{Q}\cup\eta)$. So there is a substitution v which coincides with ρ on the set $\mathcal{P}\cup\xi$ and which coincides with τ on the set $\mathcal{Q}\cup\eta$. It follows that v unifies the sets P and Q and therefore the set $P\cup Q$. Moreover, $\sigma = v|_{\xi\cup\eta}$. So, σ is an element of the right hand side of the asserted equation.

q.e.d.

Lemma 4.1.3
Let P and Q be finite sets of pairs of terms or finite sets of pairs of literals. Let ξ and η be finite sets of variables. Then there are permutations π and ϖ of variables such that

$$\mathrm{merge}(U[P,\xi], U[Q,\eta]) = U[\pi P\cup\varpi Q, \xi\cup\eta].$$

Proof: Let $\mathcal{P}$ be the set of variables occurring in P, and let $\mathcal{Q}$ be the set of variables occurring in Q. Choose π to be a permutation which coincides with the identity function on ξ and which maps $\mathcal{P}\setminus\xi$ to some set $\mathcal{V}_1 \subseteq \mathcal{V}\setminus(\xi\cup\eta)$. There certainly exists such a permutation π. Now, choose ϖ to be a permutation which coincides with the identity function on η and which maps $\mathcal{Q}\setminus\eta$ to some set $\mathcal{V}_2 \subseteq \mathcal{V} \setminus (\xi \cup \eta \cup \mathcal{V}_1)$. Again, such a permutation exists. Let $\mathcal{P}'$ be the set of variables of $\pi\mathcal{P}$ and let $\mathcal{Q}'$ be the set of variables of $\varpi\mathcal{Q}$. Then $\mathcal{P}' \subseteq \xi \cup \mathcal{V}_1$ and $\mathcal{Q}' \subseteq \eta \cup \mathcal{V}_2$. Thus $(\mathcal{P}' \cup \xi) \cap (\mathcal{Q}' \cup \eta) = \xi \cap \eta$. From Lemma 4.1.2 it follows that $\text{merge}(U[\pi P, \xi], U[\varpi Q, \eta]) = U[\pi P \cup \varpi Q, \xi \cup \eta]$. From Lemma 4.1.1 it follows that $U[P, \xi] = U[\pi P, \xi]$ and that $U[Q, \eta] = U[\varpi Q, \eta]$. So, the asserted equation holds.

q.e.d.

Lemma 4.1.4
Let P be a finite set of pairs of terms or a finite set of pairs of literals, and let ξ be a finite set of variables. Then the following two statements hold.

- $\text{forget}(U[P, \xi], x) = U[P, \xi \setminus \{x\}]$ *for* $x \in \xi$.
- $\text{rename}(U[P, \xi], f) = U[f(P), f(\xi)]$ *for each injection* $f : \xi \to \mathcal{V}$.

Now, from Lemmata 4.1.3 and 4.1.4 it follows that the three operations merge, forget, and rename, yield unifier sets as their values, as stated in the following proposition.

Proposition 4.1.1
The following three statements hold.

- *If* $\Gamma \in U_\xi$ *and* $\Delta \in U_\eta$ *then* $\text{merge}(\Gamma, \Delta) \in U_{\xi \cup \eta}$.
- *If* $\Gamma \in U_\xi$ *and* $x \in \xi$ *then* $\text{forget}(\Gamma, x) \in U_{\xi \setminus \{x\}}$.
- *If* $\Gamma \in U_\xi$ *and if* $f : \xi \to \mathcal{V}$ *is an injection then* $\text{rename}(\Gamma, f) \in U_{f(\xi)}$.

Proposition 4.1.2
The three operations merge, forget, *and* rename, *obey the following laws:*

- merge *is associative, commutative, and idempotent.*
- $\text{forget}(\text{forget}(\Gamma, x), y) = \text{forget}(\text{forget}(\Gamma, y), x)$.
- $\text{rename}(\text{rename}(\Gamma, f), g) = \text{rename}(\Gamma, gf)$.
- $\text{merge}(\text{forget}(\Gamma, x), \Delta) = \text{forget}(\text{merge}(\Gamma, \Delta), x)$, *if* $\Gamma \in U_\xi$, $\Delta \in U_\eta$, *and* $x \in \xi \setminus \eta$.

Proof: The first three assertions are trivial.

Now, consider an element σ of $\text{merge}(\text{forget}(\Gamma, x), \Delta)$. Then $\sigma|_\xi \in \text{forget}(\Gamma, x)$ and $\sigma|_\eta \in \Delta$. So there is a $\tau \in \Gamma$ such that $\tau|_{\xi \setminus \{x\}} = \sigma|_\xi$. Since $x \notin \eta$, the functions τ and $\sigma|_\eta$ coincide on the set $\xi \cap \eta$. So there is a substitution ρ on $\xi \cup \eta$ such that $\rho|_\xi = \tau$ and $\rho|_\eta = \sigma|_\eta$. So, $\rho \in \text{merge}(\Gamma, \Delta)$, and $\sigma = \rho|_{(\xi \cup \eta) \setminus \{x\}}$. It follows that $\sigma \in \text{forget}(\text{merge}(\Gamma, \Delta), x)$.

For the reverse direction, consider an element σ of $\text{forget}(\text{merge}(\Gamma, \Delta), x)$. Then there is a substitution $\rho \in \text{merge}(\Gamma, \Delta)$ such that $\sigma = \rho|_{(\xi \cup \eta) \setminus \{x\}}$. It follows that

$\rho|_\xi \in \Gamma$ and $\rho|_\eta \in \Delta$. So, $\rho|_{\xi\setminus\{x\}} \in \text{forget}(\Gamma, x)$, i.e., $\sigma|_{\xi\setminus\{x\}} \in \text{forget}(\Gamma, x)$. Moreover, from $x \notin \eta$ it follows that $\sigma|_\eta = \rho|_\eta$ and therefore $\sigma|_\eta \in \Delta$. So $\sigma \in \text{merge}(\text{forget}(\Gamma, x), \Delta)$.

q.e.d.

These properties allow us to define a merge of any finite set of unifier sets in the obvious way, and, likewise, the result of forgetting a whole set of variables rather than just one variable.

Note that, with the concept of unifier sets, the Proposition 1.3.7 in Section 1.3.3 can now be stated more simply as follows.

Proposition 4.1.3
Let H be a formula represented by a set S of clauses. Let M be a connected matrix consisting of a set of variants of clauses of S together with a set C of connections and a set F of factorization links. Assume that C together with F is spanning for M. Let $c \in M$. For each literal occurrence $L \in c$, let C_L be the set of connections not involving a literal occurrence of $c \setminus \{L\}$, and let F_L be the set of factorization links not involving a literal occurrence of $c \setminus \{L\}$. For each $L \in c$ let U_L be the unifier set of $C_L \cup F_L$ with respect to the set of variables occurring in c. Further assume that $\bigcap_{L\in c} U_L \neq \emptyset$. Then the formula H is valid.

The place of the substitution σ_L in Proposition 1.3.7 is here taken by the unifier set, say U'_L, of $C_L \cup F_L$ with respect to the set of all variables. The place of the substitution τ_L is taken by the unifier set U_L of $C_L \cup F_L$ with respect to the set of variables occurring in c. We have $U_L = \text{forget}(U'_L, \xi)$ where ξ is the set of variables not occurring in c. The place of the assumption that $\{\tau_L \mid L \in c\}$ is compatible on the set of variables occurring in c, is taken here by the assumption that $\bigcap_{L\in c} U_L \neq \emptyset$.

4.2 From Resolution to Connection Proofs

In this section we shall consider resolution refutations, and we shall see how a resolution refutation can be transformed to a complementary connected matrix, thus yielding a connection derivation of the given matrix. Of course, this is possible only at the cost of an exponential increase of the number of derivation steps in the worst case, as we already know. In fact, the number of clauses generated grows exponentially with the number of resolution steps. We shall see that this connected matrix contains repetitions which give us a hint for a more concise representation of such a connected matrix. These representations which we call *connection structures* will be formally defined in Section 4.3 and will then lead us to a proof calculus incorporating both the connection method and the resolution method.

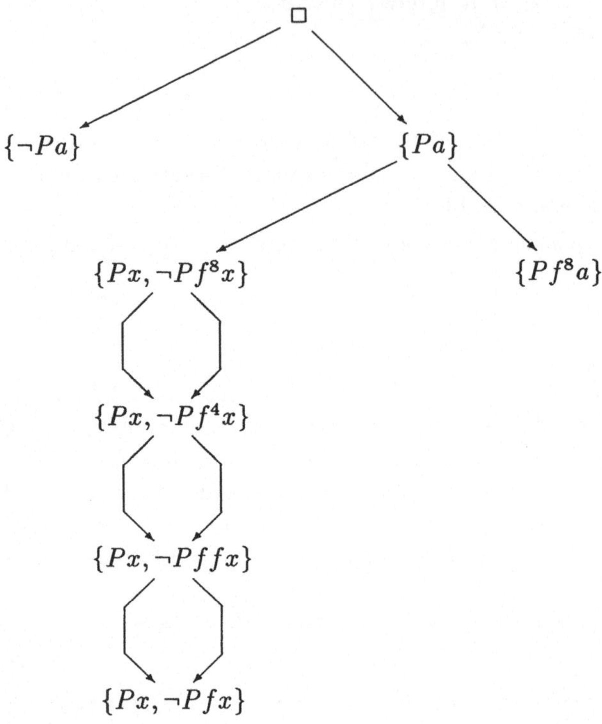

Figure 4.2: The dag representation of a resolution refutation

Let us consider again the matrix

$$\begin{array}{ccc} & Px & \\ \neg Pa & & Pf^8a \\ & \neg Pfx & \end{array}$$

The resolution proof given for this matrix in Section 2.5 has the graphical dag representation shown in Figure 4.2.

Expanding this dag we obtain the tree shown in Figure 4.3. It has ten leaves eight of which are marked with the clause $\{Px, \neg Pfx\}$. In the figure the connections resolved upon at each resolution step are drawn as thin lines between the corresponding literals. To each inner node of the tree corresponds exactly one connection in the figure, and vice versa. If n_1 and n_2 are leaves of the tree marked with clauses c_1 and c_2 and if k is one of the connections of the figure, connecting some literal of c_1 to some literal of c_2 then the node corresponding to k is the lowest node common to the branches of the tree associated with the leaves n_1 and n_2. The connections thus corresponding to the inner nodes of the tree constitute a spanning and unifiable set of connections thus resulting in a connection derivation.

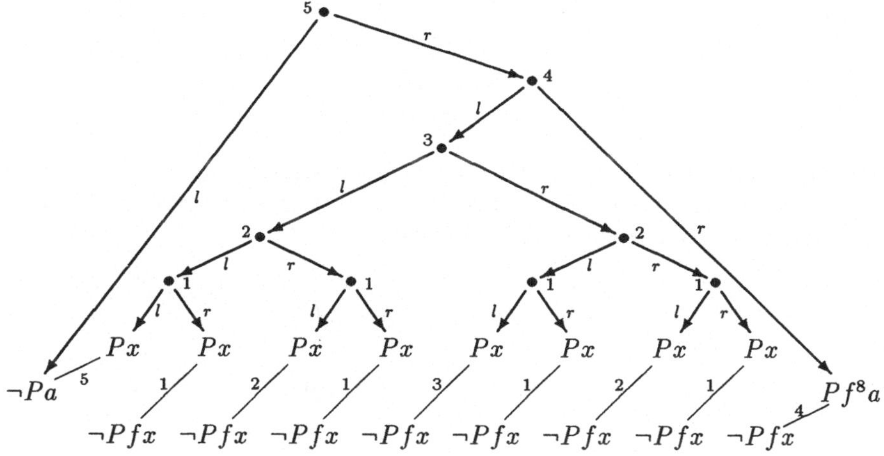

Figure 4.3: The Tree Representation of a resolution refutation

In cases where resolution steps are combined with factorization steps, we have to introduce factorization links in addition to the connections, and then the set of connections together with the set of factorization links is spanning. We shall, however, defer the consideration of factorization to Section 4.3. In our example there is no factorization.

In the case that there are no explicit or implicit factorizations, the set of connections thus generated is identical to what Peter Andrews in [And76] calls the 'mating induced' by the resolution refutation. In fact, he has arrived at the concept of matings from a study of resolution refutations and the set of connections they induce on the input clauses.

In Figure 4.3, we have attached the number n to the connections corresponding to the n-th resolution step, and also to the nodes corresponding to the resolvent of the n-th resolution step. Nodes with the same number correspond to the same node in the dag representation of the resolution derivation. In order to be able to describe the nodes of the tree, we have marked the two edges issuing from each inner node by the selector l (for "left") and by the selector r (for "right"). The choice which edge to mark with l and which one to mark with r is arbitrary but fixed.

As we see, the number of clauses and the number of connections as well as the number of nodes of the tree grows exponentially with the number of resolution steps in the worst case. However, we also see that the second and the third clause together with the connection marked with '1' between them, constitute a connected submatrix

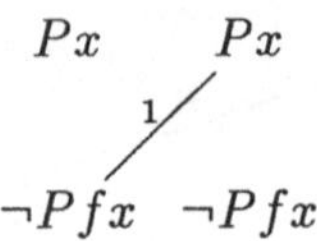

that occurs four times in the whole connected matrix. Similarly, the first two copies of this submatrix together with the connection marked with '2' between them, constitute a larger connected submatrix

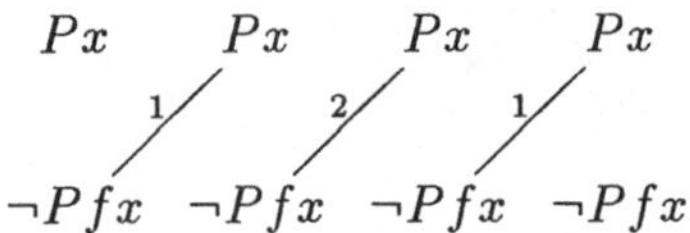

which occurs twice in the whole connected matrix. In this way we can associate a connected submatrix to each node of the tree. These connected submatrices are identical for nodes of the tree belonging to one given node of the refutation dag. So there is exactly one submatrix corresponding to each node of the resolution refutation dag, or, equivalently, to each clause occurring in the resolution refutation. In the connected matrix, we have a nested system of connected submatrices indicated by the boxes[1] below.

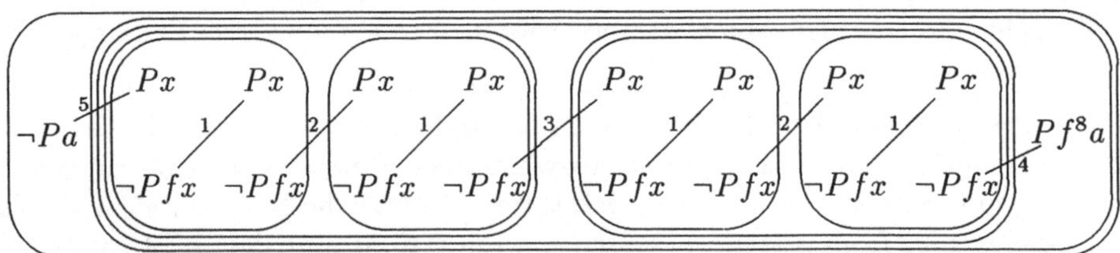

It can be seen from this picture that the number of occurrences of one single connected submatrix in the whole connected matrix may be exponential with respect to the size of the given resolution refutation. Our aim is to find a representation of the connected matrix where each of these connected submatrices is represented only once, no matter how often it occurs in the connected matrix.

Let us consider now the connected matrix of our example. Before we can try to unify its connections, we have to rename the variables in each clause such that the sets of variables are disjoint for different clauses. For each clause c, we attach the sequence of selectors leading from the root of the tree to c, as a lower index to all of its variables. Thus we get the structured matrix displayed in Figure 4.4.

[1]Of course, each clause of the matrix is itself one of these connected submatrices. For better readability, we have omitted the boxes around the clauses in the picture. Note that, within a clause of the connected matrix, no connections have been chosen.

We want to find a representation for the connections in this matrix such that all the four connections marked with '1' have a common simple representation, thus avoiding to generate explicitly four representations of connections for them. In the same sense we want the two connections marked with '2' to have a common representation.

Let us denote the node marked with 1 in the dag by n_1, the node marked with 2 in the dag by n_2, and so on. If s is a selector sequence then we denote by node(s) the node of the dag having the selector sequence s relative to the root, if such a node exists. Let Σ denote the set of all selector sequences s such that node(s) exists. For two sequences of selectors s and t let st denote the concatenation of s and t.

Let us now consider the four connections marked with 1. The set of these connections is equal to

$$\{\{\neg Pfx_{sl}, Px_{sr}\} \mid \text{node}(s) = n_1\}$$

Similarly, the set of connections marked with 2 is equal to

$$\{\{\neg Pfx_{slr}, Px_{srl}\} \mid \text{node}(s) = n_2\}$$

and the set of connections marked with 3 is equal to

$$\{\{\neg Pfx_{slrr}, Px_{srll}\} \mid \text{node}(s) = n_3\}$$

and so on.

Definition 4.2.1
Let s be a finite sequence of selectors, and let L be a literal. Then we denote by L_s the result of attaching in the literal L to each variable occurring there the selector sequence s as a lower index.

With this definition, for each of these sets of connections there are literals K and L (whose variables are not indexed), a node n of the dag, and selector sequences s_1 and s_2 such that this set of connections is equal to

$$\{\{L_{ss_1}, K_{ss_2}\} \mid s \in \Sigma \text{ and } \text{node}(s) = n\}.$$

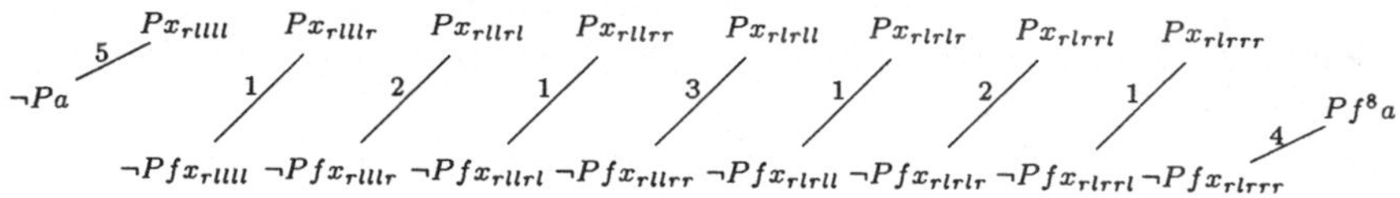

Figure 4.4: A Structured matrix

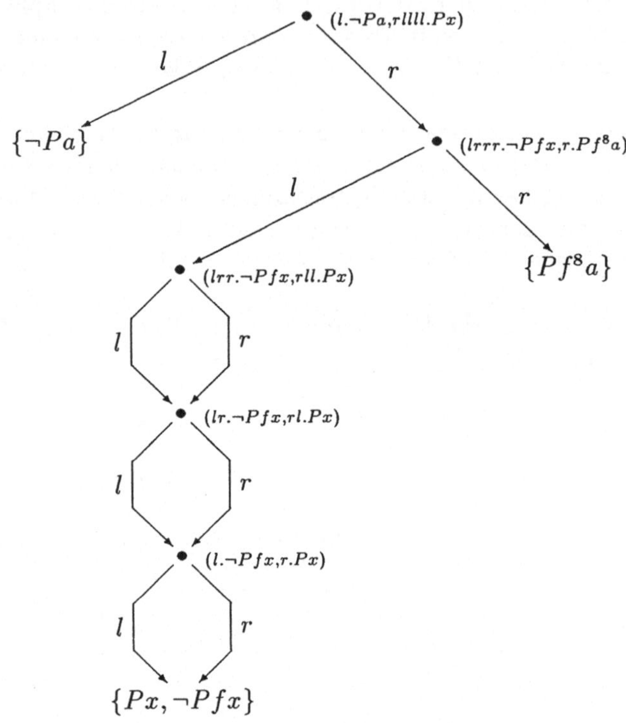

Figure 4.5: A connection structure

We denote this set of connections in the dag by attaching a label $(s_1.L, s_2.K)$ to the node n. So we obtain the dag shown in Figure 4.5. Such a dag will be called a *connection structure*. The concept of connection structures will be defined more generally in Section 4.3.

From the dag of Figure 4.5, the connected matrix can be reobtained in the following way.

- For each sequence s of selectors leading from the root to a leaf of the dag marked by some clause c, we generate a variant c_s of c by indexing the variables of c with s.

- For each $(s_1.K, s_2.L)$ marking some node n and for each selector s leading from the root to n, we draw a connection between the literal K_{ss_1} and the literal L_{ss_2}.

4.3 Connection Structures

In this section we give a formal definition of the concepts we have started to develop in Section 4.2, and we investigate some of the properties of these concepts. We begin with some basic definitions.

We shall use the word "factor" in a reflexive and transitive way in this section, i.e., a clause is a factor of itself, and a factor of a factor of a clause is again a factor of that clause. As Robinson did it in his original paper [Rob65], we shall consider a number of factorization steps followed by a resolution step, as one derivation step.[2]

When we consider resolution refutations of a matrix M in this section, we shall consider them to be represented as dags. The root of such a dag is marked with the empty clause $\Box$, and its leaves are marked with clauses of M. Such a dag will be called a *resolution dag* of M. Any inner node n which is marked with some clause c has exactly 2 parent nodes n_1 and n_2 marked with clauses c_1 and c_2. Here c is the resolvent of a factor c_1' of c_1 with a factor c_2' of c_2. We call the number $|c_1| + |c_2| - |c| - 2$ the *factorization degree* of the node n. Here $|c|$ denotes the number of literals of a clause c.

Now let M be a matrix, let C be a set of connections in M, and let F be a set of factorization links in M. Then let $\sim_F$ be the smallest (as a set of pairs of literals) equivalence relation on the set of literals of M such that $(L, K) \in \sim_F$ for all $\{L, K\} \in F$. By a *C-F-connection* we mean an unordered pair of literals $\{K, L\}$ such that there are literals K' and L' with $K' \sim_F K$ and $L' \sim_F L$ and $\{K', L'\} \in C$. We say that C together with F is *spanning* for M iff the set of all C-F-connections is spanning, i.e., iff every path through M contains a C-F-connection as a subset. We say that C together with F is *unifiable* iff there is a substitution σ unifying each connection in C as well as each factorization link in F.

Proposition 4.3.1
A matrix M is complementary iff there is a finite set M' of variants of M and a set C of connections and a set F of factorization links in M' such that C together with F is spanning for M' and unifiable.

Proof: For the proof of the direction from left to right, let M be complementary. By definition, there is a unifiable spanning set C of connections in a finite set M'

[2]Certainly, such an omission of factorization steps does not increase the size of a derivation. On the other hand, from Proposition 2.3.5 and from the fact that factorization does not increase the number of literals of a clause, it follows that reinserting the missing factorization steps leads only to a polynomial increase of the number of literals in the derivation. It may, however, increase the sizes of the literals by an exponential amount as the following example shows. Let n be a positive integer, let $x_0, \ldots, x_n$ be pairwise distinct variables, let P be an n-ary predicate symbol and f a binary function symbol, let t_j be the term $f(x_{j-1}, x_{j-1})$ for $j = 1, \ldots, n$, let K be the literal $P(x_1, \ldots, x_n)$ and L the literal $P(t_1, \ldots, t_n)$, and let c be the clause $\{K, L\}$, and let d be the clause $\{\neg K\}$. Then $(c, d, \Box)$ is a resolution refutation according to Robinson's definition. But application of a factorization step to $\{K, L\}$ yields a unit clause whose literal has exponential size. This problem can be avoided by considering terms as dags.

of variants of M. Now, just choose F to be the empty set. Then the assertion holds.

For the proof of the direction from right to left, let M' be a finite set of variants of M and let C be a set of connections and F a set of factorization links in M' such that C together with F is spanning for M' and unifiable. Then, by definition, the set of C-F-connections is spanning for M' and unifiable, and thus M is complementary.

q.e.d.

Let M be a matrix. Then an *M-dag* is a root dag D where each leaf of D is marked with a clause of M, and each edge of D is marked with a selector of some given set of selectors, any two distinct edges issuing from one node being marked by distinct selectors.

If D is an M-dag then we denote by Sel_D the set of all selector sequences leading from the root of D to some node of D, and by $\mathrm{Leafsel}_D$ we denote the set of selector sequences leading from the root of D to a leaf of D. For $s \in \mathrm{Sel}_D$ we denote by $\mathrm{node}_D(s)$ the node reached from the root of D via the selector sequence s, and for $s \in \mathrm{Leafsel}_D$ we denote by $\mathrm{clause}_D(s)$ the clause of M marking the node $\mathrm{node}_D(s)$. As in Section 4.2, we use the following notation for any selector sequence s, any literal L, and any clause c. The result of attaching s as a lower index to all variables in L or in c, is denoted by L_s or c_s, respectively.

Now assume that M is a matrix, D is an M-dag, and s, s_1, s_2 are sequences of selectors such that $s \in \mathrm{Sel}_D$ and $ss_1, ss_2 \in \mathrm{Leafsel}_D$, and assume that $K \in \mathrm{clause}_D(ss_1)$ and $L \in \mathrm{clause}_D(ss_2)$. Then the pair $(s_1.K, s_2.L)$ is said to be a *connection* for the node $\mathrm{node}_D(s)$ if K and L have opposite sign, and it is said to be a *factorization link* for the node $\mathrm{node}_D(s)$ if K and L have identical sign.

Definition 4.3.1
A *connection structure* S on a matrix M is an M-dag to each of whose nodes n is associated a set of connections $C[\mathcal{S}, n]$ for that node and a set of factorization links $F[\mathcal{S}, n]$ for that node.

The set of connections and the set of factorization links can be empty for some nodes.

Definition 4.3.2
Let $\mathcal{S}$ be a connection structure on a matrix M. Then the *full matrix* corresponding to $\mathcal{S}$ is the triple $(M_\mathcal{S}, C_\mathcal{S}, F_\mathcal{S})$ where

- $M_\mathcal{S} \overset{\text{def}}{=} \{c_s \mid s \in \mathrm{Leafsel}_D \;\text{ and }\; c = \mathrm{node}_D(s)\}$
- $C_\mathcal{S} \overset{\text{def}}{=} \{(K_{ss_1}, L_{ss_2}) \mid s \in \mathrm{Sel} \;\text{ and }\; (s_1.K, s_2.L) \in C[\mathcal{S}, \mathrm{node}(s)]\}$
- $F_\mathcal{S} \overset{\text{def}}{=} \{(K_{ss_1}, L_{ss_2}) \mid s \in \mathrm{Sel} \;\text{ and }\; (s_1.K, s_2.L) \in F[\mathcal{S}, \mathrm{node}(s)]\}$.

Here D denotes the underlying dag of $\mathcal{S}$.

We say that a connection structure $\mathcal{S}$ is *spanning* or *unifiable* iff the set $C_{\mathcal{S}}$ of connections together with the set $F_{\mathcal{S}}$ of factorization links is spanning for $M_{\mathcal{S}}$ or unifiable, respectively.

Proposition 4.3.2
A matrix M is complementary iff there is a spanning and unifiable connection structure for M.

Proof: If $\mathcal{S}$ is a spanning and unifiable connection structure for M then $C_{\mathcal{S}}$ together with $F_{\mathcal{S}}$ is spanning for the finite set $M_{\mathcal{S}}$ of variants of clauses of M and unifiable, and so M is complementary.

If, on the other hand, M is complementary, then there is a unifiable and spanning set of connections and factorization links, in a finite set of variants of clauses of M. Let $c^1,\ldots,c^r$ be those clauses. By renaming the variables, we get a set C of connections and a set F of factorization links on the clauses $c^1_{s_1},\ldots,c^r_{s_r}$ where $s_1,\ldots,s_r$ are pairwise distinct selectors. Then C together with F is spanning and unifiable. Now we take, as a dag D for our connection structure, a tree whose nodes are its root n which is not marked, and its leaves $n_1,\ldots,n_r$ which are marked with the clauses $c^1,\ldots,c^r$. The edges of the tree are $(n,n_1),\ldots,(n,n_r)$, and they are marked with the selectors $s_1,\ldots,s_r$. Now let $C[\mathcal{S},n]$ be the set of $(s_i.K, s_j.L)$ such that $K \in c^i$ and $L \in c^j$ and that $\{K_{s_i}, L_{s_j}\} \in C$. The set $F[\mathcal{S},n]$ of factorization links is defined similarly. For each node n_i different from n, let $C[\mathcal{S},n_i] = \emptyset$ and $F[\mathcal{S},n_i] = \emptyset$. The so defined connection structure $\mathcal{S}$ on M is spanning and unifiable.

q.e.d.

We see from this proof that the proposition remains true when we require $F_{\mathcal{S}}$ to be empty, i.e., as in the connection method, we do not need factorization in the connection structures in order to assure completeness. It has been introduced only for improving efficiency and for allowing a direct simulation of resolution which is guaranteed to exist by the following proposition.

Proposition 4.3.3
Let M be a matrix and let Δ be a resolution dag of M. Then there is a spanning and unifiable connection structure $\mathcal{S}$ such that the underlying dag D of $\mathcal{S}$ is isomorphic to Δ and for each inner node of D the associated set of connections is a singleton set and the cardinality of the associated set of factorization links is equal to the factorization degree of the corresponding node of Δ.

Proof: First, we attach to each edge in the resolution dag Δ a selector as a label in such a way that two distinct edges issuing from one node always have two distinct selectors as their labels. This is certainly possible. Now, for each node n marked with some clause c in the resolution dag Δ, we define

- the (empty or singleton) set $C[\mathcal{S},n]$ of connections for n

- the set $F[\mathcal{S}, n]$ of factorization links for n
- the object $\mathcal{L}[L, n]$ for each literal L of c

by induction on the difference of the depth of the dag Δ and the depth of the node n in Δ. The object $\mathcal{L}[L, n]$ will be a string of the form $r.L'$, where L' is a literal in one of the input clauses of Δ such that L is a descendant of L', and where r is the selector sequence leading from L to L'. The inductive definition is as follows.

If n is a leaf then $C[\mathcal{S}, n]$ and $F[\mathcal{S}, n]$ are the empty set and $\mathcal{L}[L, n] \stackrel{\text{def}}{=} \epsilon.L$ where ϵ is the empty sequence of selectors. Otherwise let n_1 and n_2 be the parent nodes of n and let s_1 and s_2 be the selectors leading from n to n_1 and n_2, respectively, and let c_1 and c_2 be the clauses marking the nodes n_1 and n_2.

If $K \in c_1$ and $L \in c_2$ are the literals resolved upon at the resolution step corresponding to node n, and if $\mathcal{L}[K, n_1] = r.K'$ and $\mathcal{L}[L, n_2] = s.L'$, then $C[\mathcal{S}, n] \stackrel{\text{def}}{=} \{(s_1 r.K', s_2 s.L')\}$.

Now, for the definition of $\mathcal{L}[L, n]$ and $F[\mathcal{S}, n]$, let us assume that $L \in c$, and let d_1 and d_2 be the sets of literals of c_1 and c_2, respectively, of which L is an immediate descendant, i.e., $d_i = \{K \in c_i \mid \sigma_i K = L\}$ for $i = 1, 2$ where σ_1 and σ_2 are the substitutions applied to c_1 and c_2, respectively, in that derivation step. If no factorization is involved then $|d_1| + |d_2| = 1$. Now let

$$A_i \stackrel{\text{def}}{=} \{s_i s.K' \mid \text{ there is a } K \in d_i \text{ such that } \mathcal{L}[K, n_i] = s.K'\},$$

and let $A \stackrel{\text{def}}{=} A_1 \cup A_2$. Now we choose B_L to be an arbitrary minimal set of pairs of elements of A having as its reflexive-symmetric-transitive closure the set $A \times A$, and we choose an arbitrary element of A as our object $\mathcal{L}[L, n]$. We have now defined a set B_L for every literal L of c. Let $F[\mathcal{S}, n]$ be the union of these sets.

The so defined connection structure $\mathcal{S}$ has the properties required.

q.e.d.

Finally a remark for the reader who is familiar with Peter Andrews' work [And76]. Let $\mathcal{C}$ be the set of $C_\mathcal{S}$-$F_\mathcal{S}$-connections in the proof of Proposition 4.3.3. Then $\mathcal{C}$ is equal to what Andrews calls the 'mating induced' by the resolution refutation Δ. More precisely, if $\mathcal{R}$ is the tree expansion of the refutation Δ with variables renamed as described above then $\mathcal{C}$ is the mating induced by $\mathcal{R}$.

In order to see this, let us have a look at the mating induced by some arbitrary resolution refutation $\mathcal{R}$. It is defined as the set of all pairs of literal-occurrences in the input clauses which have descendants which are resolved with each other in $\mathcal{R}$ (see [And76]). Now, consider the set of literal-occurrences in the input clauses of $\mathcal{R}$ together with the mating induced by $\mathcal{R}$, as an undirected graph $\mathcal{G}$. Then, to each resolution step R corresponds exactly one component of $\mathcal{G}$, i.e., one maximal connected subgraph of $\mathcal{G}$, and vice versa. This component consists of those literal-occurrences L in the input clauses of $\mathcal{R}$ such that one of the two literals resolved upon at step R is a descendant of L. In each component, every positive literal is mated with every negative literal.

Now consider, for the proof of Proposition 4.3.3, the resolution refutation $\mathcal{R}$ defined above. The set $C_{\mathcal{S}}$ of connections of the full matrix corresponding to $\mathcal{S}$, contains exactly one connection from each component of $\mathcal{G}$. The equivalence relation $\sim_{F_{\mathcal{S}}}$ generated by the factorization links of the full matrix consists of the unordered pairs $\{L, K\}$ where L and K are literal-occurrences which are both negative or both positive and which both occur in the same component of $\mathcal{G}$. So, indeed, $\mathcal{C}$ is equal to the mating induced by $\mathcal{R}$.

Note, however, a few differences between connection structures and matings. The most important difference in terms of lengths of shortest proofs is that the connection structure method uses a dag representation of the resolution refutation whereas matings are derived from their tree representation and therefore are exponentially longer. This can be seen from the example given in Section 2.5. Secondly, in a connection structure, each connection (and each factorization link) is associated with one node of the dag thus giving the set of connections a hierarchical structure which allows to exploit the possibilities opened by the dag representation of the refutation as opposed to the tree representation. The third difference is that in a connection structure a resolution step is always represented by only one connection and a number of factorization links which is less than the sum of the cardinalities of the parent clauses, whereas in a mating a resolution step is represented by a number of connections which may be as bad as quadratic even with respect to the length of a tree resolution refutation (and exponential with respect to the length of a dag resolution refutation).

4.4 The Connection Structure Calculus

In the last section we have seen that resolution refutations can be represented by connection structures, and that also for each connection proof we can find a unifiable and spanning connection structure on the given matrix. In order for connection structures to be useful for theorem proving, we must have a calculus in addition which allows us to generate connection structures and to decide quickly whether they are spanning and unifiable.

Definition 4.4.1
An *annotated connection structure* $\mathcal{A}$ is a connection structure $\mathcal{S}_{\mathcal{A}}$ together with two functions $\Gamma_{\mathcal{A}}$ and $\mathcal{P}_{\mathcal{A}}$ yielding for each node n of the corresponding dag a unifier set $\Gamma_{\mathcal{A}}(n)$ and a set $\mathcal{P}_{\mathcal{A}}(n)$ of partial paths in the full matrix M_n corresponding to the substructure $\mathcal{S}_n$ of $\mathcal{S}_{\mathcal{A}}$ at node n.

Definition 4.4.2
An annotated connection structure $\mathcal{A}$ is said to be *correct* if, for each node n, the following two propositions hold.

- Let $\mathcal{L}$ be the set of all connections and factorization links in the full matrix (M_n, C_n, F_n) corresponding to $\mathcal{S}_n$. Then $\Gamma_{\mathcal{A}}(n)$ is the unifier set of $\mathcal{L}$ with respect to some set of variables.

- Every path through the full matrix M_n that does not contain a C_n-F_n-connection, has some element of $\mathcal{P}_{\mathcal{A}}(n)$ as a subset.

Proposition 4.4.1
Let $\mathcal{A}$ be a correct annotated connection structure for some matrix M. Let n be the root of $\mathcal{A}$ and assume that $\Gamma_{\mathcal{A}}(n) \neq \emptyset$ and $\mathcal{P}_{\mathcal{A}}(n) = \emptyset$. Then M is complementary.

Proof: Let (M', C, F) be the full matrix corresponding to $\mathcal{S}_{\mathcal{A}}$, and let $\mathcal{L}$ be the set of all C-F-connections. Then M' is a finite set of variants of clauses of M, and $\mathcal{L}$ is a set of connections in M'. By definition, the non-empty set $\Gamma_{\mathcal{A}}(n)$ is the unifier set of $C \cup F$ and therefore also of $\mathcal{L}$ with respect to some set of variables. It follows that $\mathcal{L}$ is unifiable. Moreover, by definition, every path through M' that does not contain a connection of $\mathcal{L}$, has some element of $\mathcal{P}_{\mathcal{A}}(n)$, i.e., of $\emptyset$, as a subset. So every path through M' contains a connection of $\mathcal{L}$. In other words, $\mathcal{L}$ is spanning for M'. We know now that $\mathcal{L}$ is a unifiable spanning set of connections in M'. By definition, it follows that M is complementary.

q.e.d.

Now, for a spanning and unifiable connection structure $\mathcal{S}$ on some matrix M, sets $\Gamma_{\mathcal{A}}(n)$ and $\mathcal{P}_{\mathcal{A}}(n)$ for each node n can be directly calculated so that a correct annotated connection structure $\mathcal{A}$ is obtained for the matrix M. This is done by defining these two sets for each leaf and then inheriting them from the leaves to the immediate predecessors of the leaves, and so on, until the root is reached, as it is described below.

First we introduce a few notations. The variables occurring in the matrix M are denoted by x,y,.... We introduce a new variable x_s for each variable x occurring in M and for each selector sequence s. If s is a selector or a sequence of selectors then let $f[s]$ be the injection on the set of these variables which is defined by $f[s](x_r) \stackrel{\text{def}}{=} x_{sr}$ for all variables x occurring in M and for all selector sequences r.

Definition 4.4.3
Let $\mathcal{A}$ be an annotated connection structure and let n be a node of $\mathcal{A}$ whose successor nodes $n_1,\ldots,n_k$ are reached from n via the selectors $s_1,\ldots,s_k$, respectively. Further let C be the set of connections for n and let F be the set of factorization links for n. Then the *unifier set inherited* from the successors of n is defined to be the set

$$\text{merge}(\text{rename}(\Gamma_{\mathcal{A}}(n_1), f[s_1]), \ldots, \text{rename}(\Gamma_{\mathcal{A}}(n_k), f[s_k]), C, F).$$

The *path set inherited* from the successors of n is defined to be the set

$$\{f[s_1](p_1) \cup \ldots \cup f[s_k](p_k) \mid p_1 \in \mathcal{P}_{\mathcal{A}}(n_1), \ldots, p_k \in \mathcal{P}_{\mathcal{A}}(n_k)\}$$

Definition 4.4.4
We say an annotated connection structure $\mathcal{B}$ is obtained from an annotated connection structure $\mathcal{A}$ by a step of *introduction of new links*, $\mathcal{A} \vdash_{\text{links}} \mathcal{B}$, if the following conditions hold.

1. $\mathcal{B}$ has exactly the same nodes as $\mathcal{A}$ plus one additional node n to which is associated some set C of connections and some set F of factorization links.

2. $\mathcal{B}$ has exactly the same edges as $\mathcal{A}$ plus some number k of additional edges from node n to nodes $n_1, \ldots, n_k$, marked with selectors $s_1, \ldots, s_k$.

3. All edges of $\mathcal{A}$ are marked the same way in $\mathcal{B}$ as in $\mathcal{A}$. The same holds for the nodes and for their marks, associated sets of connections and factorization links, and for their annotations.

4. If Γ is the unifier set inherited from the successors of n then

$$\Gamma_{\mathcal{A}}(n) = \text{forget}(\Gamma, \xi)$$

   where ξ is a set of variables neither occurring in $C[\mathcal{S}_{\mathcal{A}}, n]$ nor in $F[\mathcal{S}_{\mathcal{A}}, n]$.

5. For every element p of the path set inherited from the successors of n, the following condition holds. If p does not contain a C-F-connection then there is a set $p' \in \mathcal{P}_{\mathcal{A}}(n)$ such that $p' \subseteq p$.

Definition 4.4.5
We say that an annotated connection structure $\mathcal{B}$ is obtained from an annotated connection structure $\mathcal{A}$ on a matrix M by a step of *introduction of a new clause*, $\mathcal{A} \vdash_{\text{clause}} \mathcal{B}$, if the following conditions hold.

1. $\mathcal{B}$ is identical to $\mathcal{A}$ except that it has an additional isolated node n, i.e., a node of indegree and outdegree zero.

2. n is marked by some clause c of M, and $\Gamma_{\mathcal{A}}(n) = U_\xi$ where ξ is the set of variables occurring in c, and $\mathcal{P}_{\mathcal{A}}(n) = \{\{L\} \mid L \in c\}$.

Definition 4.4.6
By $\vdash$ we denote the transitive-reflexive closure of $\vdash_{\text{link}} \cup \vdash_{\text{clause}}$, and we say that $\mathcal{B}$ is *derivable from* $\mathcal{A}$ if $\mathcal{A} \vdash \mathcal{B}$. In the case that $\mathcal{A}$ is the annotated connection structure with zero nodes we say that $\mathcal{B}$ is *derivable*, $\vdash \mathcal{B}$. In this case we also say that the underlying matrix M is *derivable in the connection structure calculus.*

In the following proposition we mean by a *linear dag* a dag that has one branch $(n_1, \ldots, n_k)$ such that all nodes not on this branch are leaves and every immediate successor of a node $n_j (j = 1, \ldots, k-1)$ is a leaf or identical to the node n_{j+1}.

Proposition 4.4.2
A matrix M is complementary iff it is derivable. Moreover there exists a real number α such that, for all nonnegative integers κ and n and for each resolution refutation of size κ and length (i.e., number of derivation steps) n for some matrix M, there is a derivation in the connection structure calculus which has a size $\leq \alpha\kappa$ and length n.

For each connection proof of a matrix M, there is a derivation in the connection structure calculus whose number of derivation steps equals the number of choices of connections and factorization links done in the connection proof plus the number of variants of clauses in the connection proof. Moreover, the dag of the correct annotated connection structure is a linear dag. The simulation is a p-simulation.

Proof: In both cases we have a step by step simulation in the connection structure calculus. In the case of the simulation of resolution we have to choose $\mathcal{P}_{\mathcal{A}}(n)$ to be the set $\{\{L\} \mid L \in c\}$ where c is the clause corresponding to the node n, and the set ξ of variables in the definition of $\Gamma_{\mathcal{A}}(n)$ is chosen as the set of variables occurring in c, i.e., we forget all variables not occurring in c. The dag underlying the derivation thus constructed in the connection structure calculus directly mirrors the given resolution refutation, and the dag underlying the obtained connection structure is actually isomorphic to the given resolution refutation dag. This is not surprising from the way we have arrived at connection structures.

Now, the simulation of a connection proof could be done by simulating it in resolution and then simulating the resulting resolution proof in the connection structure calculus. But this is not possible with linear input resolution, and so the tree that we obtain in this way is not linear. We shall use a more direct simulation here. In each step of the construction of our linear dag the dag will always be a root dag, except for possible isolated nodes due to clauses that have been introduced but that do not have any connections yet in the constructed connection structure. Let $\mathcal{D}$ be the given derivation in the connection method. Our construction will be such that, at each instant of time, the full matrix corresponding to the connection structure constructed so far, will be exactly the structured matrix constructed up to that instant of time in the connection calculus. So the simulation will be step by step and it will preserve the temporal order. Our first step is an introduction of a new clause, namely of the starting clause of $\mathcal{D}$.

Corresponding to each extension step of $\mathcal{D}$ involving a new variant of a clause c, we make a step of introduction of a new clause, namely of c, in the connection structure calculus. In addition, we make a step of an introduction of new links for each extension step, namely of the connections corresponding to those which are used in the extension in $\mathcal{D}$ at that step. The new root n of the dag introduced in this step has as its immediate successors the old root n_1 of the dag and the node n_2 marked with the clause extended to. As the set ξ of variables forgotten in this step we choose the empty set. In the connection calculus a concept of forgetting a variable does not exist. For the path set $\mathcal{P}_{\mathcal{A}}(n)$ we take the set of all paths in the full matrix M constructed so far which have the form $(K_1, \ldots, K_k, L)$ where $(K_1, \ldots, K_r)$ is the active path and $0 \leq k \leq r$ and L is an unsolved subgoal obeying the following condition. If $k < r$ then L is in the same clause as K_{k+1}, and if $k = r$ then L is in the clause extended to.

Corresponding to each factorization step of $\mathcal{D}$ we make a step of an introduction of new links, namely of the factorization links involved in that factorization step. The new root n of the dag has exactly one successor node, namely the old root n_1 of the dag. Again, the set ξ of forgotten variables is empty. The path set $\mathcal{P}_{\mathcal{A}}(n)$ is again defined to be the set of those partial paths which are initial segments of the active path extended into an unsolved subgoal of the next clause of the matrix.

Whereas in the simulation of resolution all elements of $\mathcal{P}_{\mathcal{A}}(n)$ are singleton sets, we have more complex partial paths here. But note that the length of a partial path is at most equal to the number of proof steps and that the cardinality of each path set is at most equal to the product of the number of proof steps and the maximal cardinality of the input clauses. The current most general unifier is the same as in the connection method. So, in fact, the simulation is a p-simulation.

q.e.d.

Actually, the derivation in the connection structure calculus just constructed is almost identical to the given derivation in the connection calculus. The only difference is that in the connection derivation, the path set is encoded more efficiently by structure sharing. But such an encoding can be done just as well in the connection structure calculus.

Theorem 4.4.1
The connection structure calculus can p-simulate resolution and the connection calculus.

The relevance of forgetting of variables in the connection structure calculus becomes apparent with our example

$$\begin{matrix} & Px & \\ \neg Pa & & Pf^8a \\ & \neg Pfx & \end{matrix}$$

Consider the connection structure given for this matrix in Figure 4.5. Let n_0 be the root of the dag in Figure 4.5, let n_1 be its right successor, let n_2 be the left successor of n_1, let n_3 be the successor of n_2, and let n_4 be the successor of n_3. Then without forgetting of variables, the most general unifier corresponding to n_4 is $\{x_r \leftarrow fx_l\}$, the mgu corresponding to n_3 is $\{x_{lr} \leftarrow fx_{ll}, x_{rl} \leftarrow ffx_{ll}, x_{rr} \leftarrow fffx_{ll}\}$, the mgu corresponding to n_2 is $\{x_{llr} \leftarrow fx_{lll}, x_{lrl} \leftarrow f^2x_{lll}, x_{lrr} \leftarrow f^3x_{lll}, x_{rll} \leftarrow f^4x_{lll}, x_{rlr} \leftarrow f^5x_{lll}, x_{rrl} \leftarrow f^6x_{lll}, x_{rrr} \leftarrow f^7x_{lll}\}$, the mgu corresponding to n_1 is $\{x_{llll} \leftarrow a, x_{lllr} \leftarrow fa, x_{llrl} \leftarrow f^2a, x_{llrr} \leftarrow f^3a, x_{lrll} \leftarrow f^4a, x_{lrlr} \leftarrow f^5a, x_{lrrl} \leftarrow f^6a, x_{lrrr} \leftarrow f^7a\}$, and the mgu corresponding to n_0 is $\{x_{rllll} \leftarrow a, x_{rlllr} \leftarrow fa, x_{rllrl} \leftarrow f^2a, x_{rllrr} \leftarrow f^3a, x_{rlrll} \leftarrow f^4a, x_{rlrlr} \leftarrow f^5a, x_{rlrrl} \leftarrow f^6a, x_{rlrrr} \leftarrow f^7a\}$. With forgetting of variables, the mgu for n_3 simplifies to $\{x_{rr} \leftarrow fffx_{ll}\}$, the mgu for n_2 simplifies to $\{x_{rrr} \leftarrow f^7x_{lll}\}$, the mgu for n_1 simplifies to $\{x_{llll} \leftarrow a\}$, and the mgu for n_0 simplifies to the empty substitution. It is obvious that when we take the example

$$\begin{matrix} & Px_1 \ldots x_m & \\ \neg Pa_1 \ldots a_m & & \neg Pa_{\pi^{-1}(1)} \ldots a_{\pi^{-1}(m)} \\ & \neg Px_{\pi(1)} \ldots x_{\pi(m)} & \end{matrix}$$

of Section 2.5 instead of this one, we obtain a polynomial size of the derivation with forgetting of variables, whereas the size of the derivation without forgetting of variables is superpolynomial.

4.5 Splitting with Connection Structures

We shall indicate briefly in this section how splitting can be done in the connection structure calculus.

Let us assume that the hypotheses of Proposition 1.3.7 are fulfilled. Let M_0 be the matrix consisting of all clauses of M except the clause c, and let us consider on M_0 the set of all links that are also links of M. Then there is an annotated connection structure $\mathcal{A}$ corresponding to M_0 with these links. Let m be its root. Now, let $L_1, \ldots, L_k$ be the literals of c, and let l and r be two selectors. We introduce $k+1$ new nodes $n_0, \ldots, n_k$ where n_0 is added as a new leaf to $\mathcal{A}$ and marked with the clause c. For $j = 1, \ldots, k$, the node n_j is added as a new node with immediate successors m and n_{j-1} where the corresponding selectors are l and r. For each connection or factorization link $\kappa = \{K, L_j\}$ between a literal K of M_0 and the literal L_j let $\tilde{\kappa} \stackrel{\text{def}}{=} (ls.K, r^j.L_j)$ where s is the selector sequence corresponding to the literal K in the connection structure $\mathcal{A}$ and r^j denotes the selector sequence $r \ldots r$ with j occurrences of r. Then we mark the node n_j with the set of all connections of the form $\tilde{\kappa}$ and with the set of all factorization links of the form $\tilde{\kappa}$. Let $\mathcal{S}$ be the connection structure thus obtained. Then the full matrix corresponding to $\mathcal{S}$ is obtained from M by taking k variants of the submatrix M_0 together with the clause c and by choosing connections and factorization links exactly as it is done for splitting in the connection method (see Section 1.3.3).

Now, to obtain an annotated connection structure $\mathcal{A}$, we have to describe how the set of partial paths $\mathcal{P}_{\mathcal{A}}(n_j)$ and the unifier set $\Gamma_{\mathcal{A}}(n_j)$ is obtained for each of the new nodes n_j $(j = 0, \ldots, k)$.

The set of partial paths is obtained in a way quite analogous to the way it is chosen in the simulation of the connection calculus (without splitting) in the connection structure calculus. The only difference is that the paths containing L_j go through the j-th variant of the submatrix M_0 whereas for the connection calculus without splitting there is only one variant of M_0.

Now, the unifier sets $\Gamma_{\mathcal{A}}(n_j)$ are uniquely determined by giving the set ξ_j of variables such that $\Gamma_{\mathcal{A}}(n_j) \in U_{\xi_j}$. This follows from the definition of the concept of an annotated connection structure and from the fact that we have already defined the set of connections and factorization links for each node. So, for the definition of $\mathcal{A}$ it suffices to give the set ξ_j of variables for each node n_j. We define ξ_j to be the set of variables $f[r^j](x)$ such that x is a variable occurring in the clause c.

We have to show that the so defined unifier sets are not empty. Now let $j \in \{0, \ldots, k\}$, and let $\mathcal{S}_j$ be the subdag of $\mathcal{S}$ at the node n_j. Let Γ_j be the set of unifiers of the set of links in the full matrix corresponding to $\mathcal{S}_j$. Since splitting is only allowed if this set of links is unifiable, the set Γ_j is not empty. Now, the unifier set $\Gamma_{\mathcal{A}}(n_j)$ is the set of all restrictions of elements of Γ_j to the set ξ_j. So $\Gamma_{\mathcal{A}}(n_j)$ is not empty.

Thus we obtain a correct annotated connection structure for M. Note that the unifier sets thus constructed are obtained from the unifier set corresponding to M_0 and from the unifier sets for the newly chosen links by the operations 'merge' and 'rename'. If there are variables that will not be needed any more in the further course of the derivation then we can, in addition, apply the forget operation.

Instead of adding $k+1$ nodes, we can add just two nodes, one leaf marked with the clause c, and the new root of the dag. Then we also get a correct annotated connection structure. The outdegree of the new root of a splitting step is $k+1$. So we do not have a binary dag any more.

The following proposition cannot be proved in the mathematical sense since we have not given a formal definition of the concept of splitting by actually formulating a calculus that could be called 'the connection calculus with splitting'. Rather, we have to appeal to the description given above of a way to do splitting in the connection structure calculus.

Proposition 4.5.1
Let F be a formula in clausal form. Let $n \in \mathbf{N}$ and let there be a proof in the connection calculus with splitting in which the number of occurrences of chosen links equals n. Then there are proofs with at most n steps in the connection structure calculus and in the resolution calculus. The connection structure calculus can p-simulate the connection calculus with splitting.

4.6 Extended Definitional Calculi

The extension rule can be added to any calculus of clausal form logic, and any calculus of clausal form logic can be carried over to full first order logic by means of the transformation to definitional form. In particular, this is also true of the connection structure calculus.

Definition 4.6.1
By a *derivation* of a formula F of full first order predicate logic in the *extended definitional connection structure calculus* we mean a pair $(\mathcal{E}, \mathcal{D})$ where $\mathcal{E}$ is a derivation of a set S of clauses from def_F by extension steps, and $\mathcal{D}$ is a derivation of S in the connection structure calculus. If, in addition, $\mathcal{E}$ is the empty sequence (i.e., if there are no extension steps) then we say that $(\mathcal{E}, \mathcal{D})$ is a *derivation* of F in the *definitional connection structure calculus*.

From the results in Chapters 3 and 4 we obtain the following theorem

Theorem 4.6.1
The definitional connection structure calculus can p-simulate the cut-free sequent calculus. The extended definitional connection structure calculus can p-simulate the (full) sequent calculus, the natural deduction calculus, and the Frege-Hilbert calculus.

Similarly, we can introduce the (extended) definitional connection calculus. It has already been mentioned that the connection method is directly applicable not only to formulas in clausal form, but to arbitrary formulas of full first order logic.

The calculi for the non-clausal connection method look somewhat more complex than the corresponding calculi for clause form logic (see [Bib87]). Now, the connection calculus for full first order logic can be p-simulated in the definitional (clausal form) connection calculus, as has been done in an actual implementation in [Ede85a]. The correspondence between the non-clausal connection method and the definitional (clause form) connection method is so close that such an implementation can be viewed both, as a direct implementation of the non-clausal connection method, or as an implementation of the definitional (clause form) connection method. So, the definitional form can serve as a key to understand the non-clausal connection method, and, maybe, also as a key to understand some other non-clausal theorem proving methods.

Conclusion

We have seen that the repeated use of a formula as a lemma is an essential part of concise proofs of formulas. Proof calculi which do not have this feature (such as the tableau calculus and the simpler versions of the connection method) cannot p-simulate typical calculi which have this feature (such as resolution). Nevertheless, the connection method is suited for theorem proving in two ways even in cases where its simplest version is inefficient. First, it is a good tool for a theoretical analysis of proofs in other calculi such as resolution, and second, it can therefore be used to construct on top of it a concise meta-language for expressing possibly long connection proofs efficiently. The connection structure calculus can be viewed as such a meta-language where the transformation from an expression in the meta-language to the expression it denotes is exponential. But in an actual derivation, this transformation never need to be done.

Many questions concerning the p-simulatability of the considered calculi have not been treated here. Even some very simple results have not been mentioned, for example, that the sequent calculus[3], the natural deduction calculus, and the Frege-Hilbert calculi, can p-simulate resolution in the sense that any resolution derivation of a clause c from a clause set S can be transformed to a derivation of the clause formula corresponding to c from the clause formulas corresponding to the elements of S. This p-simulatability of resolution does not imply, however, that also definitional resolution can be p-simulated in the sequent calculus, in natural deduction, or in a Frege-Hilbert calculus. This would be the converse of some of the results of Chapter 3. It would be interesting to investigate the question whether the converses of any of the results of Chapter 3 hold.

There is one question in Chapter 4 that has been left open and that would be interesting to answer. Namely, it has been shown there that the connection calculus and resolution can be p-simulated in the connection structure calculus. Moreover, the connection structure calculus was obtained by mapping a resolution refutation (at exponential cost) to a derivation in the connection calculus, and then removing the exponential amount of redundancies again to obtain a resolution refutation dag with certain additional structures on top of it. But the connection structure calculus is a little bit more general than this. One reason for this is that the connection structure calculus is not restricted to binary dags as resolution is (since each resolvent has only two parent clauses in resolution). The other, and probably

[3]Note that this is not possible in the sequent calculus without cut. In fact, resolution is very closely connected to the cut rule.

more important, way in which the connection structure calculus is more general than resolution, is that the connection structure calculus allows arbitrary path sets. In contrast, any connection structure obtained from a resolution refutation by the simulation process, contains only one-element paths. Remember that the path set corresponding to a resolvent or input clause c occurring in a resolution refutation, is the set of all one-element paths $\{L\}$ such that $L \in c$. From all this, it is not clear whether resolution can p-simulate the connection structure calculus. My guess is that it can. We have seen that, despite of the fact that Bibel's connection calculus deals with paths of arbitrary lengths, it can be p-simulated in resolution (which allows only one-element paths). So, my guess is that a similar construction would allow to p-simulate the connection structure calculus in resolution.

Nothing has been said in this book about the relative complexity of the definitional connection method or the extended definitional connection method in relation to the other calculi. It seems that such examples as the formulas $Pa \wedge \forall(Px \rightarrow Pfx) \rightarrow Pf^n x$ which we considered in Section 2.5, have short proofs in the extended definitional connection calculus. So, it may well be that the extended definitional connection calculus can also simulate the sequent calculus, the natural deduction calculus, and the Frege-Hilbert calculus. But these questions have to be left to future research.

There are some questions still open concerning the simulation of definitions in extended definitional resolution (Section 3.8). We have allowed definitions only for predicates, not for functions. In the Begriffsschrift, definitions are allowed for both. The extension rule would have to be generalized still more to include functions. It would then be natural to include equality. Equality could be treated either by introducing equality axioms, or by paramodulation. Another possible generalization would be the substitution rule stating roughly the following. If a formula has been derived then any formula obtained from it by replacing each n-ary predicate symbol or function symbol with an n-ary nominal form (in the sense of Kurt Schütte [Sch77]) of the appropriate type, can be derived from it. Frege used this rule in derivations in his Begriffsschrift, although he did not explicitly state it as a rule. In propositional logic, the renaming of variables is a special case of the substitution rule. Renaming of propositional variables together with extended definitional resolution allows to p-simulate the substitution rule in propositional logic.

Since Gerhard Gentzen has introduced his sequent calculus for intuitionistic logic as well as for classical logic, the question arises whether intuitionistic logic can also be p-simulated by some sort of intuitionistic extended definitional resolution. There is one problem, however, with this idea. The formulas D_G defined in Section 2.2 cannot be simplified as much as in classical logic. For the propositional connectives, the problem is not too severe. The concept of a clause would have to be modified. We would have to allow formulas of the form

$$\forall x_1 \ldots \forall x_k (I_1 \wedge \cdots \wedge I_m \rightarrow A_1 \vee \cdots \vee A_n)$$

as clause formulas where $x_1, \ldots, x_k$ are the variables in the matrix, each A_j is an atomic formula, and each I_j is an implication of the form $B_j \rightarrow C_j$ where

B_j and C_j are atomic formulas. We could define an intuitionistic resolution for such formulas. We could hope that this would again yield p-simulatability of the sequent calculus, natural deduction, and the Frege-Hilbert calculus in extended definitional resolution. But this works only for propositional logic. In predicate logic, skolemization can destroy the soundness of the calculus, as the following example shows. Let F be the formula

$$\forall x(Px \vee \neg Px) \rightarrow \exists x Px \vee \neg \exists x Px,$$

and let G be the following formula obtained from it by skolemizing the last occurrence of $\exists x Px$.

$$\forall x(Px \vee \neg Px) \rightarrow \exists x Px \vee \neg Pc.$$

Then F is not valid, but G is valid, in intuitionistic logic.

There are some calculi that have not been treated here, for example the rewrite methods for theorem proving by Jieh Hsiang and Nachum Dershowitz [HD83]. The propositional connectives they use are $+$ (the exclusive 'or') and $\cdot$ (the connective $\wedge$). Quantifier-free formulas are then terms in the term algebra over the function symbols, the predicate symbols, and the connectives $+$ and $\cdot$. The set of such terms is a boolean ring with respect to $+$ and $\cdot$ with a zero element 0 (falsum) and a unit element 1 (verum). Terms are considered to be equal if they can be shown equal by the laws of the boolean ring. Universally quantified formulas are considered as equations of the form $t = 0$. Such equations can be transformed to rewrite rules. By applying the techniques of overlapping by Knuth and Bendix [KB70], a calculus is obtained for deriving new rewrite rules from old ones, or new equations from old ones. If, finally, $1 = 0$ has been derived then the given formula has been refuted.

Now, every equation can be written in the form $t = 0$. If $t = 0$ is the result of an application of the rule of overlapping to two equations $r = 0$ and $s = 0$ then t is a linear combination $p\sigma r + q\tau s$ of a substitution instance σr of r and of a substitution instance τs of s. Let us say, t is obtained from r and s by application of rule R. Then each resolution step can be obtained also by an application of rule R. If we stop identifying terms that can be shown to be equivalent by the basic rules of the boolean ring and if we add these rules to the rule R instead, then we obtain a proof system. This proof system can p-simulate resolution. The converse is not obvious since there are formulas such as $P_1 + \cdots + P_n$ in the boolean ring that cannot be translated to clausal form at polynomial cost preserving semantic equivalence. It would be interesting to know whether a proof in this proof system of a formula that is already in clausal form can be p-simulated by resolution and whether, for a proof of an arbitrary formula F in this system, there is a resolution proof of polynomial size for def_F.

Bibliography

[AHM87] Arnon Avron, Furio A. Honsell, and Ian A. Mason. Using typed lambda calculus to implement formal systems on a machine. Technical Report ECS-LFCS-87-31, Department of Computer Science, University of Edinburgh, July 1987.

[And76] Peter B. Andrews. Refutations by matings. *IEEE Transactions on Computers*, C-25(8):801–807, August 1976.

[And81] Peter Andrews. Theorem proving via general matings. *Journal of the ACM*, 28(2):193–214, April 1981.

[BB87] K. H. Bläsius and H. J. Bürckert, editors. *Deduktionssysteme, Automatisierung des logischen Denkens.* Oldenbourg Verlag, München, Wien, 1987.

[BdlT89] Thierry Boy de la Tour. A locally optimal transformation into clause form using partial formula renaming. Rapport de recherche, Laboratoire d'Informatique Fondamentale et d'Intelligence Artificielle, Institut IMAG, Saint Martin d'Hères, France, 1989.

[BEF83] Wolfgang Bibel, Elmar Eder, and Bertram Fronhöfer. Towards an advanced implementation of the connection method. In Alan Bundy, editor, *Proceedings of the Eighth International Joint Conference on Artificial Intelligence ijcai-83*, pages 920–922, Los Altos, California, August 1983. William Kaufmann, Inc.

[BEK+87] Stephan Bayerl, Elmar Eder, Franz Kurfeß, Reinhold Letz, and Johannes Schumann. An implementation of a prolog-like theorem prover based on the connection method. In Ph. Jorrand and V. Sgurev, editors, *Artificial Intelligence II—Methodology, Systems, Applications (AIMSA'86), Varna, Bulgaria*, pages 29–36, Amsterdam, 1987. Association pour la Promotion de l'Informatique Avancée (APIA), North-Holland.

[Bet55] Evert W. Beth. Semantic entailment and formal derivability. *Medедlingen der Koninklijke Nederlandse Akademie van Wetenschappen*, 18(13):309–342, 1955.

[Bet59] Evert W. Beth. *The Foundations of Mathematics.* North-Holland, Amsterdam, 1959.

[Bib83] Wolfgang Bibel. Matings in matrices. *Communications of the ACM*, 26(11):844–852, November 1983.

[Bib87] Wolfgang Bibel. *Automated Theorem Proving.* Artificial Intelligence. Vieweg, Braunschweig/Wiesbaden, second edition, 1987.

[Bib89] Wolfgang Bibel. Short proofs of the pigeonhole formulas based on the connection method. *Journal of Automated Reasoning*, 1989. To appear.

[Bur87] R. M. Burstall. Research in interactive theorem proving at Edinburgh University. In *Proceedings of the 20th IBM Computer Science Symposium, Shizuoka, Japan*, 1987.

[Bus87] Samuel R. Buss. Polynomial size proofs of the propositional pigeonhole principle. *The Journal of Symbolic Logic*, 52(4):916–927, December 1987.

[C+86] Robert L. Constable et al. *Implementing Mathematics with the NuPRL Proof Development System.* Prentice-Hall, Englewood Cliffs, NJ, 1986.

[Chu36] Alonzo Church. An unsolvable problem of elementary number theory. *American Journal of Mathematics*, 58:345–363, 1936. Reprinted in [Dav65], pp. 88–107.

[CL73] Chang and Lee. *Symbolic Logic and Mechanical Theorem Proving.* Academic Press, New York, 1973.

[Coo71] Stephen A. Cook. The complexity of theorem-proving procedures. In *Proceedings of Third Annual ACM Symposium on Theory of Computing, Shaker Heights, Ohio*, pages 151–158, 1971.

[Coo76] Stephen A. Cook. A short proof of the pigeonhole principle using extended resolution. *ACM SIGACT News*, 8:28–32, Oct.–Dec. 1976.

[CR74] Stephen A. Cook and Robert A. Reckhow. On the lengths of proofs in the propositional calculus. In *Proceedings of Sixth Annual ACM Symposium on Theory of Computing, Seattle, Washington*, pages 135–148, 1974. Corrections for this paper are in Sigact News, July 1974, Vol. 6, No. 3, pages 15–22.

[CR79] Stephen A. Cook and Robert A. Reckhow. The relative efficiency of propositional proof systems. *The Journal of Symbolic Logic*, 44(1):36–50, March 1979.

[Dav65] Martin Davis. *The Undecidable. Basic Papers on Undecidable Propositions, Unsolvable Problems and Computable Functions. Gödel, Church, Turing, Rosser, Kleene, Post.* Raven Press, Hewlett, New York, 1965.

[dB80] Nicolas G. de Bruijn. A survey of the project automath. In J. P. Seldin and J. R. Hindley, editors, *To H. B. Curry: Essays in Combinatory Logic, Lambda Calculus, and Formalism*, pages 589–606. Academic Press, 1980.

[Ede85a] Elmar Eder. An implementation of a theorem prover based on the connection method. In W. Bibel and B. Petkoff, editors, *Artificial Intelligence, Methodology, Systems, Applications (AIMSA'84)*, pages 121–128, Amsterdam, New York, Oxford, 1985. European Coordinating Committee for Artificial Intelligence, North-Holland.

[Ede85b] Elmar Eder. Properties of substitutions and unifications. *Journal of Symbolic Computation*, 1:31–46, 1985.

[Ede87] Elmar Eder. Theorem proving and the connection method—the state of the art. *Rendiconti del Seminario Matematico, Università e Politecnico di Torino, Logic and Computer Science (1986)*, pages 93–114, 1987. Fascicolo Speciale.

[Ede89] Elmar Eder. A comparison of the resolution calculus and the connection method, and a new calculus generalizing both methods. In E. Börger, H. Kleine Büning, and M. M. Richter, editors, *CSL'88, 2nd Workshop on Computer Science Logic, Duisburg, FRG, October 1988, Proceedings, Lecture Notes in Computer Science 385*, pages 80–98, Berlin, Heidelberg, New York, 1989. Springer-Verlag.

[Ede91] Elmar Eder. Consolution and its relation with resolution. In *Proceedings of the 12th International Joint Conference on Artificial Intelligence (IJCAI-91), Sydney*, pages 132–136. Morgan Kaufmann, August 1991.

[Fre79] Gottlob Frege. Begriffsschrift, eine der arithmetischen nachgebildete Formelsprache des reinen Denkens. Halle, 1879. Engl. Transl. in [Hei67], pp. 1–82.

[Gen35] Gerhard Gentzen. Untersuchungen über das logische Schließen. *Mathematische Zeitschrift*, 39:176–210, 405–431, 1935. Engl. transl. in [Sza69], pp. 68–131.

[GNOP82] S. Greenbaum, A. Nagasaka, P. O'Rorke, and D. A. Plaisted. Comparison of natural deduction and locking resolution implementations. In D. Loveland, editor, *Proceedings of the 6th Conference of Automated Deduction*, pages 159–171. Springer Lecture Notes in Computer Science 138, 1982.

[Göd30] Kurt Gödel. Die Vollständigkeit der Axiome des logischen Funktionenkalküls. *Monatshefte für Mathematik und Physik*, 37:349–360, 1930. Engl. transl. in [Hei67], pp. 582–591.

[Göd31] Kurt Gödel. Über formal unentscheidbare Sätze der Principia Mathematica und verwandter Systeme. *Monatshefte für Mathematik und Physik*, 38:173–198, 1931. Engl. transl. in [Hei67], pp. 596–616 and [Dav65], pp. 4–38.

[HA72] D. Hilbert and W. Ackermann. *Grundzüge der theoretischen Logik.* Die Grundlehren der mathematischen Wissenschaften in Einzeldarstellungen. Springer-Verlag, Berlin, Heidelberg, New York, 1972.

[Hak85] Armin Haken. The intractability of resolution. *Theoretical Computer Science*, 39:297–308, 1985.

[HB68] D. Hilbert and P. Bernays. *Grundlagen der Mathematik.* Grundlehren der mathematischen Wissenschaften in Einzeldarstellungen. Springer, Berlin, Heidelberg, New York, 1968.

[HD83] Jieh Hsiang and Nachum Dershowitz. Rewrite methods for clausal and non-clausal theorem proving. In Joseph Díaz, editor, *Automata, Languages and Programming, 10th Colloquium, ICALP 83, Barcelona, Spain, July 1983*, pages 331–346, Berlin, Heidelberg, New York, Tokio, 1983. Springer-Verlag.

[Hei67] Jean van Heijenoort. *From Frege to Gödel. A Source Book in Mathematical Logic, 1879–1931.* Source Books in the History of the Sciences. Harvard University Press, Cambridge, Massachusetts, 1967.

[Her76] H. Hermes. *Einführung in die mathematische Logik.* Teubner, Stuttgart, 1976. English translation: Introduction to Mathematical Logic, Springer-Verlag 1973.

[HHP87] Robert Harper, Furio Honsell, and Gordon Plotkin. A framework for defining logics. In *Proceedings of the Symposium on Logic in Computer Science, Ithaca, New York, June 1987*, pages 194–204. IEEE Computer Society Press, 1987.

[Hue76] Gérard Huet. *Résolution d'équations dans des langages d'ordre* $1, 2, \ldots, \omega$. PhD thesis, L'Université de Paris VII, 1976.

[KB70] Donald E. Knuth and Peter B. Bendix. Simple word problems in universal algebras. In John Leech, editor, *Computational Problems in Abstract Algebra, Oxford 1967*, pages 263–297, Oxford, 1970. Science Research Council, Pergamon Press.

[Let91] Reinhold Letz. *First-Order Calculi and Proof Procedures for Automated Deduction.* PhD thesis, Technische Hochschule Darmstadt, Germany, 1991.

[Lin88] Peter A. Lindsay. A survey of mechanical support for formal reasoning. *Software Engineering Journal (UK)*, 3(1):3–27, January 1988.

[Lov78] Donald W. Loveland. *Automated Theorem Proving: A Logical Basis.* North-Holland, 1978.

[Man77] Yu. I. Manin. *A Course in Mathematical Logic.* Graduate Texts in Mathematics 53. Springer-Verlag, New York, Heidelberg, Berlin, 1977.

[MF86] Dale Miller and Amy Felty. An integration of resolution and natural deduction theorem proving. In *Proceedings of the aaai-86, fifth national conference on artificial intelligence, august 1986, philadelphia, pa*, pages 198–202, Los Altos, Cal., 1986. american association for artificial intelligence, Morgan Kaufmann publishers, Inc. vol. 1, science.

[MM82] A. Martelli and U. Montanari. An efficient unification algorithm. *ACM Transactions on Programming Languages and Systems*, 4:258–281, 1982.

[MR89] Neil V. Murray and Erik Rosenthal. Short proofs of the pigeonhole formulas using path dissolution. Submitted to *Discrete Applied Mathematics*, 1989.

[PG86] David Plaisted and Steven Greenbaum. A structure-preserving clause form translation. *Journal of Symbolic Computation*, 2:293–304, 1986.

[Poo84] D. L. Poole. Making clausal theorem provers non-clausal. In *Proc. CSCSI/SCEIO Conf.*, pages 124–125, London, Ontario, 1984.

[PW78] M. S. Paterson and M. N. Wegman. Linear unification. *J. Comp. Syst. Sci.*, 16:158–167, 1978.

[Rec76] Robert Allen Reckhow. *On the Lengths of Proofs in the Propositional Calculus.* PhD thesis, University of Toronto, 1976.

[Ric78] Michael M. Richter. *Logikkalküle.* Studienbücher: Informatik, Leitfäden der angewandten Mathematik und Mechanik, Bd. 43. Teubner, Stuttgart, 1978.

[Rob65] J. A. Robinson. A machine-oriented logic based on the resolution principle. *Journal of the ACM*, 12:23–41, 1965.

[Sch77] Kurt Schütte. *Proof Theory.* Grundlehren der mathematischen Wissenschaften 225. Springer-Verlag, Berlin, Heidelberg, New York, 1977. Transl. from German.

[Sho67] Joseph R. Shoenfield. *Mathematical Logic.* Addison-Wesley, Reading, Massachusetts, 1967.

[Sko20] Thoralf Skolem. Logisch-kombinatorische Untersuchungen über die Erfüllbarkeit oder Beweisbarkeit mathematischer Sätze nebst einem Theoreme über dichte Mengen. *Videnskapsselskapets skrifter, I. Matematisk-naturvidenskabelig klasse*, (4), 1920.

[Smu71] Raymond M. Smullyan. *First-Order Logic.* Ergebnisse der Mathematik und ihrer Grenzgebiete. Springer-Verlag, Berlin, Heidelberg, New York, 1971.

[SVvK84] Wolfgang Stegmüller and Matthias Varga von Kibéd. *Strukturtypen der Logik*, volume 3 of *Probleme und Resultate der Wissenschaftstheorie und Analytischen Philosophie.* Springer-Verlag, Berlin, Heidelberg, New York, Tokyo, 1984.

[SW83] Jörg Siekmann and Graham Wrightson, editors. *Automation of Reasoning, Classical Papers on Computational Logic.* Springer, Berlin, 1983.

[Sza69] M. E. Szabo. *The Collected Papers of Gerhard Gentzen.* Studies in Logic and the Foundations of Mathematics. North-Holland, Amsterdam, 1969.

[Tse70] G. S. Tseitin. On the complexity of derivations in propositional calculus. In A. O. Slisenko, editor, *Studies in Constructive Mathematics and Mathematical Logic*, pages 115–125. Consultants Bureau, New York, 1970. part II.

[Urq87] Alasdair Urquhart. Hard examples for resolution. *Journal of the ACM*, 34(1):209–219, 1987.

[WOLB84] L. Wos, R. Overbeek, E. Lusk, and J. Boyle. *Automated Reasoning: Introduction and Applications.* Prenctice-Hall, Eaglewood Cliffs, 1984.

[WR13] Alfred North Whitehead and Bertrand Russel. *Principia Mathematica.* Cambridge University Press, Cambridge (England), 1910/13.

Index

Parallelism in Logic

Its Potential for Performance and Program Development

von Franz Kurfeß

1991. XII, 299 pp. (Artificial Intelligence, ed. by Wolfgang Bibel and Walther von Hahn) Softcover.
ISBN 3-528-05163-9

The potential of parallelism in logic reaches far beyond the exploitation of AND- and OR-parallelism usually found in attempts to parallelize PROLOG. This book discusses parallelism in logic and its exploitation on parallel architectures. A variety of categories of parallelism is discussed with respect to different levels of a logical formula and different ways to evaluate it. As an outcome of these investigations ot is shown that modularity allows structuring of logic programs and meta-evaluation can be used to control the evaluation process on a parallel system. This combination yields a consistent programming framework with a wide scope. Finally, the suitability of a specific evaluation mechanism for parallel architectures is investigated.

Vieweg Publishing · P.O. Box 58 29 · D-6200 Wiesbaden

vieweg